CAMBRIDGE GUIDE TO THE WEATHER

Ross Reynolds

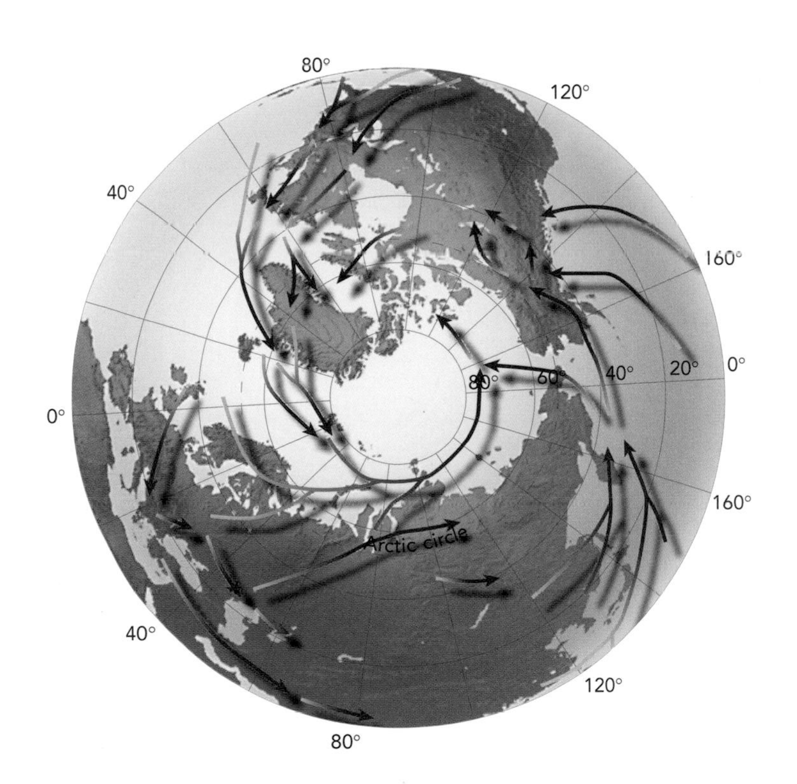

Preface

Meteorology is perhaps the prime example of a science that is exposed to public scrutiny, every day, globally. The weather forecast you see on television or hear on the radio is the end product of a massive, continuing international effort to observe, exchange, collate, and process a wide variety of observations of the atmosphere as quickly as possible in order to provide a timely, accurate prediction. People may joke, but the fact is that the quality of weather forecasts – out to a week or more ahead, has improved significantly over the last few decades. This is true on a global scale, and also for severe events such as hurricanes, typhoons and the much shorter-lived tornadoes.

This positive change has occurred because the observational network has, in the main, improved around the world. It has happened because incredibly high capacity, high-speed computers are used in the world's major weather forecast centres both to represent many important factors more precisely and to process extremely large numbers of observations very rapidly indeed. In addition, they facilitate more sophisticated research into every aspect of weather science.

Most of the research undertaken both in weather services and universities is aimed at improving our understanding of the complexities of the atmosphere, with the ultimate aim, for example, of improving forecasts of day-to-day weather or the global climate some decades or even a century ahead. The text and figures in this book have been produced to illustrate some of these strands of modern meteorological science.

The subject has always attracted many enthusiasts not lucky enough to be able to work in the subject professionally. Hopefully, some readers will take up the opportunity of joining their national weather 'Society' and, if the possibility exists, to consult some or all of the Web sites quoted in the book.

My thanks are due to Caroline Rayner, Steve Scanlan, Laura Walker, Simon Roberts and Kara Turner for their friendly and efficient help in the preparation and production of this volume.

To Mum, Dad, Carole and Lucy

The illustrations have been prepared by Stefan Chabluk and Julian Baker. Where possible, both Imperial and metric measurements have been used. A conversion table is given on the inside back cover to aid conversion where this has not been possible.

PUBLISHED BY THE PRESS SYNDICATE OF THE UNIVERSITY OF CAMBRIDGE
The Pitt Building, Trumpington Street, Cambridge CB2 1RP, United Kingdom

CAMBRIDGE UNIVERSITY PRESS
The Edinburgh Building, Shaftesbury Road, Cambridge, CB2 2RU, United Kingdom
http://www.cup.cam.ac.uk
40 West 20th Street, New York, NY 10011–4211, USA
http://www.cup.org
10 Stamford Road, Oakleigh, Melbourne 3166, Australia

First published by Cambridge University Press in 2000

Printed in China

This edition only for sale in the United States of America and Canada

ISBN 0 521 77489 6 paperback

Introduction

The weather influences all of us either directly or indirectly. In rare circumstances it can threaten our very lives with tornados, thick fog, flood-producing rain, lightning and more. Since most of us live and work in urban areas and travel by modern private or public transport, we are less aware of, and less affected by, the weather than we would have been decades ago.

I well remember my elementary school teacher saying that the weather is always a useful medium for striking up a conversation with a stranger. That was in the late Fifties. Since then, especially during the last decade or so, scientists have made us all aware of changes in our atmosphere that are potentially of great consequence to our lives. Nowadays, just about all of us are at least conscious of predicted future changes in our atmosphere that are linked to the way we live.

Amongst other things, this book outlines a grounding in meteorology for those who want to gain a basic understanding of what makes the weather 'tick', and to appreciate what lies behind some of today's great atmospheric environmental issues. If you are interested, for example, in being more involved in taking your own observations, obtaining advice about using weather science in schools, linking up with others interested in the atmospheric environment and receiving magazines and newsletters, there are learned societies you can contact, all with regional centers. All welcome foreign members too.

In the United States, the American Meteorological Society encourages anyone interested in weather and climate to join. Similar societies exist in the UK, Australia, New Zealand and Canada. They can be contacted by post or via the Web. In addition, in the UK and some other European countries, the "Climatological Observers' Link" exists to promote the collection and exchange of weather observations by anyone interested to do so.

USA: Executive Secretary
American Meteorological Society
45 Beacon Street
BOSTON
MA 021083693
http://ametsoc.org/AMS/amshomepage.cfm

Canada: Executive Secretary
Canadian Meteorological &
Oceanographic Society
112(c)150 Louis Pasteur
OTTAWA
Ontario K1N 6N5
http://www.cmos.ca

UK: Executive Secretary
Royal Meteorological Society
104 Oxford Road
READING RG1 7LL
http://itu.reading.ac.uk/rms/rms.html

Dr R Brugge
Climatological Observers' Link
16 Wooton Way
MAIDENHEAD SL6 4QU
http://www.met.rdg.ac.uk/~brugge/col.html

Australia: Australian Meteorological &
Oceanographic Society
Administrative Office
POB 654E
MELBOURNE
Victoria 3001
http://atmos.es.mq.edu.au/AMOS/index.html

New Zealand: Secretary
Meteorological Society of New Zealand
POB 6523
Te Aro
WELLINGTON
http://metsoc.rsnz.govt.nz/

In a few places within the book I have quoted Web addresses for major sites that provide up-to-date details of various topics within weather and climate. These open up a vast resource for those readers who are connected to the World Wide Web.

chapter one 1

The Atmosphere

Before looking at the weather in detail, it is necessary to set the scene by explaining some basic facts about the atmosphere's chemical make-up, its structure and layering, and what drives it to stir on a global scale. This basic knowledge will help you understand subsequent chapters.

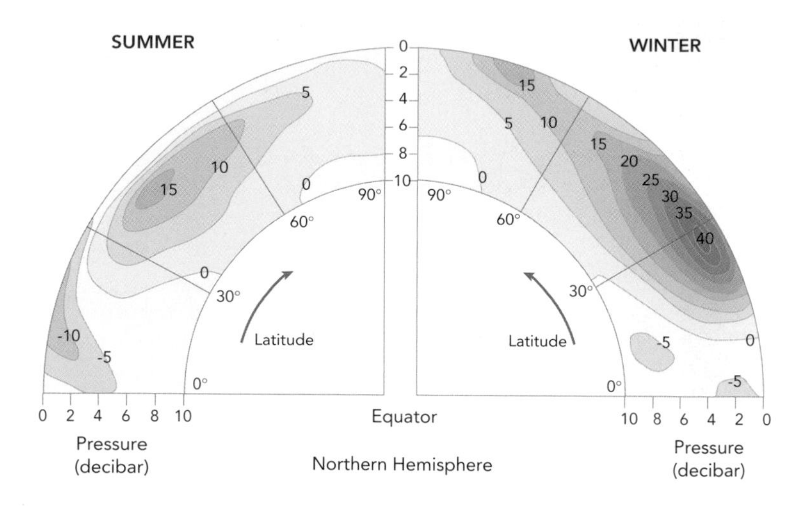

Average summer (left) and winter (right) northern hemisphere westerly (positive)/easterly (negative) wind speed (m/sec)

Observations of the weather taken at sites scattered across the Earth's surface and up through the atmosphere provide the essential information for producing an accurate global picture of its patterns. Measurements taken throughout the year enable weather scientists to study the nature and extent of the major seasonal changes that occur in these patterns, at the surface and at different levels within the atmosphere.

Zonal mean is the average of all the separate values within a latitude band

Meteorologists produce a wide variety of different weather maps of variables, such as average monthly or seasonal variations at the surface or through some level within the atmosphere. Additionally, they use north–south vertical cross-sections that may stretch from pole to pole and depict, for example, the annual mean temperature from the surface up to 30 km [18 miles]. A common technique with this type of section is to average all the values within a latitude band to calculate an average value at each level for the observation points scattered through the band. This is repeated, say, for every ten-degree latitude band, allowing a sequence of 'zonal mean' values to be calculated.

The average change in temperature from the surface up through the atmosphere is generated from vast numbers of thermal observations taken across all latitudes and through the depth of the atmosphere. This summary of an immense data set is the basis for the definition of the layers of the Earth's atmosphere.

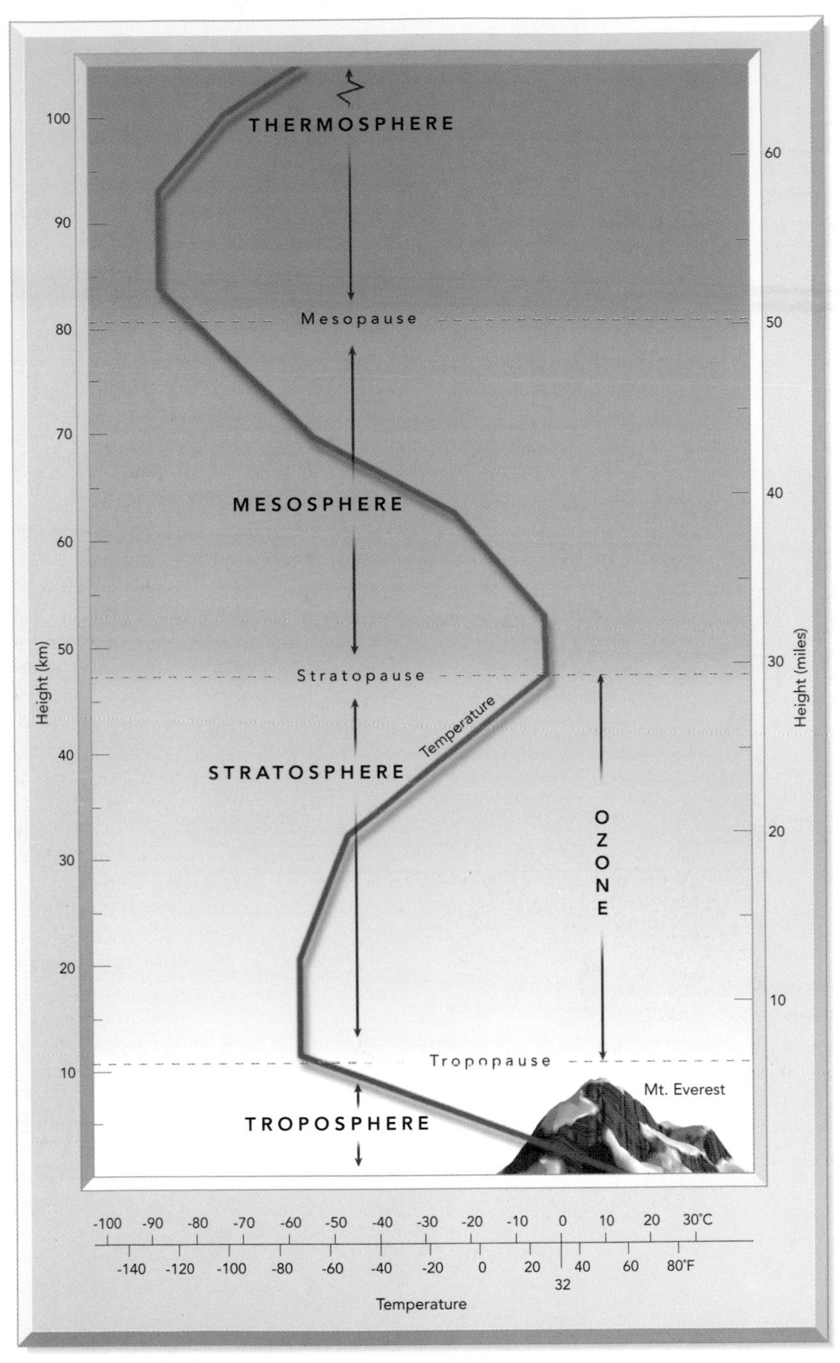

Average thermal profile of the atmosphere

Make-up

The primitive atmosphere of the Earth may have been produced by gases escaping from within the planet as it was warmed by such processes as radioactive decay. Since that time, the composition of the air has changed dramatically, partly due to the evolution of life. More recently, industrialized human society has had a major impact on the make-up of the atmosphere, the potential consequences of which are very much to the fore today (Chapter 9).

Dry air

We know that the gases that comprise the "dry" atmosphere occur in fixed proportions up to about 100 km [60 miles] above sea level. This well-mixed layer is known as the turbosphere, and within it the mixing is carried out by large-scale weather systems and much smaller-scale turbulence.

This region is capped by the turbopause, above which the atmosphere is characterized by layers composed of individual gases separated out according to their molecular weight, shown in the table below. The heavier gases occur at the lower levels of the upper atmosphere.

Water and aerosol

"Dry" air is so called because it omits a very significant additional component – water vapor. This is not included because, unlike the foregoing constituents, it is highly variable in concentration and resides at lower levels, mainly within the first few kilometers of the atmosphere because it originates at the Earth's surface.

In addition to these gaseous constituents, the air within the lower levels of the atmosphere (the troposphere) contains solid and liquid water in the form of ice, water droplets, clouds, and precipitation. Also, very small particles,

Component (Gas)	**Molecular weight**	**Fraction of total molecules**
Nitrogen	28.02	0.7808
Oxygen	32.00	0.2095
Argon	39.94	0.0093
Water vapor	18.02	0 to 0.04
Carbon dioxide	44.01	340 parts per million
Neon	20.18	18 parts per million
Helium	4.00	5 parts per million
Krypton	83.70	1 part per million
Hydrogen	2.02	0.5 parts per million
Ozone	48.00	0 to 12 parts per million

known as aerosol, occur in the same layer; they comprise a suspension of solid and liquid particles with very low settling velocities and their diameters range from about 1/10,000,000,000 of a meter to 1/100,000 of a meter!

Pressure

Pressure is related to the weight of the air above the point at which the measurement is taken. This means that it must be highest at the Earth's surface and decrease continually upward through the atmosphere until a tenuous ill-defined edge is reached at its outer limit.

The air is compressed under its own weight, so its density also decreases with height. The link between the air's density, pressure, and elevation above sea level is summarized in the table on page 10, which uses prominent natural and man-made features as examples.

The annual global mean-sea-level pressure is 1013.2 mbar, which relates to an air density of 1.23 kg/m^3. Note how even at the top of the Telecom Tower in London, UK, the pressure is, on average, some 23 mbar lower than at sea level. At the top of the Empire State Building in New York, USA, it is

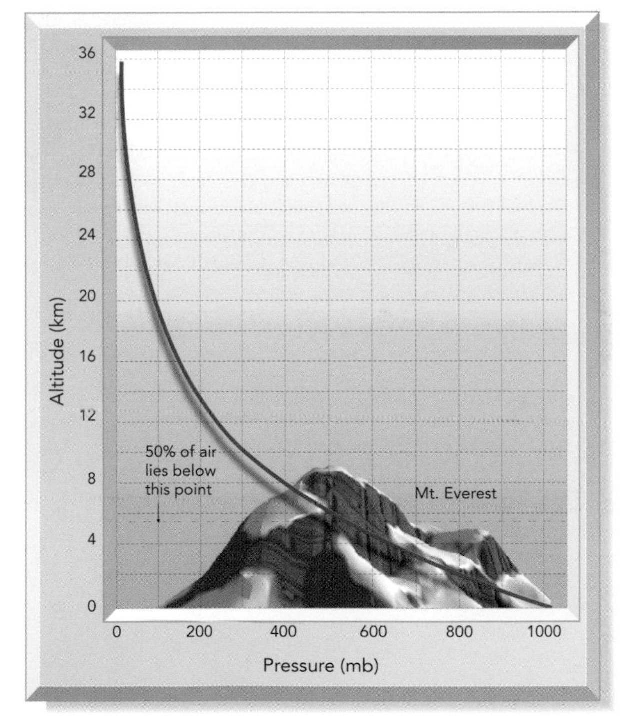

Atmospheric pressure change with height

typically 53 mbar lower! Air at the harborside in New York is about 3% denser than the air at the top of the Empire State. Taller buildings have even lower atmospheric pressure and air density figures, but nowhere near as low as those found at the peaks of mountains.

At the top of Ben Nevis, the UK's highest peak, the air density is just over 1.0 kg/m^3 and the pressure 860 mbar. Going up through the highest peaks in Australia, Western Europe, North America and the rest of the world leads to thinner and thinner air, down to a density of 0.48 kg/m^3 at the top of Mount Everest, where the average pressure is 315 mbar.

It is crucial that aircraft cabins are pressurized given the even thinner air at the height they fly.

The troposphere

The troposphere is the lowest layer of the atmosphere. It is characterized by temperatures that, on average, decrease with height, and by the presence of almost all the atmosphere's clouds and weather. The prefix "tropos" is Greek for "turn", and it was chosen because the layer is generally well mixed by vertical circulations of the air, which vary in depth and vigor. This type of air motion is a hallmark of the troposphere, although it does not occur everywhere all of the time.

Because the depth of the overturning motions is related to the intensity of surface heating, on average the layer is deepest in the tropics and becomes shallower toward the poles. There is a seasonal variation outside the lowest latitudes

Approximate values of air density and pressure at different heights

Site	**Height** (amsl*) (m)	**Air density** (kg/m^3)	**Pressure** (mbar)
Mean-sea-level	0	1.23	1013.2
Telecom Tower, London, UK	189	1.20	990.0
Sky Tower Auckland, New Zealand	328	1.18	970.0
Empire State Building, New York, USA	448	1.17	960.0
Petronas Towers, Kuala Lumpur, Malaysia	452	1.17	958.0
Ben Nevis, UK	1343	1.07	860.0
Mount Kosciuszko, Australia	2280	0.95	780.0
Mont Blanc, France	4810	0.75	550.0
Mount McKinley (Denali), USA	6194	0.65	460.0
Mount Everest, Nepal	8848	0.48	315.0
Cruising 747 'jumbo' jet	11,000	0.36	225.0
Cruising Concorde	18,300	0.12	70.0

* amsl = above mean-sea-level

such that the troposphere is deepest in summertime. The average lapse rate of temperature – the rate at which it falls with height – is almost 6°C/km [17°F/mile], but values can vary greatly with time and space.

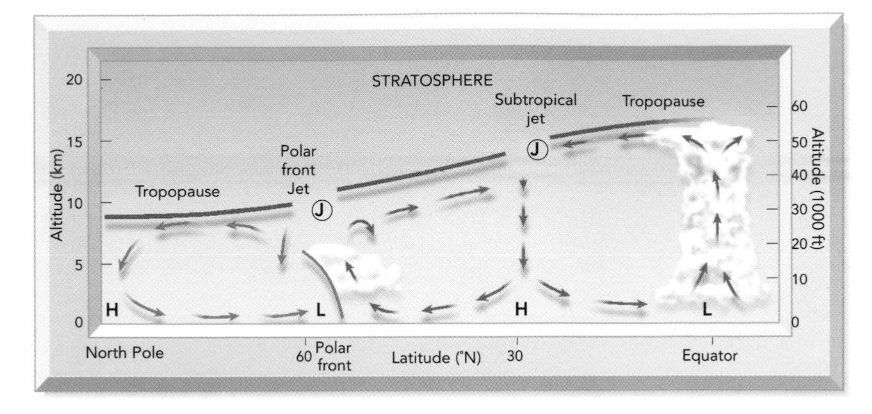

Average wintertime depth of the troposphere, location of the major westerly jetstreams and surface pressure centers

When air ascends, it cools at a rate that depends on whether it is "dry" (i.e. without clouds) or "saturated" (cloudy). Conversely, when it descends it warms at a rate that depends on the same factors. Ascending air moves into steadily reducing pressure, which causes it to expand and consequently cool. Descending air is compressed as it subsides gradually into higher pressure and thus is warmer. So the up-and-down motions that typify the troposphere are associated with cooler air aloft and warmer air below.

Something like 80% of the mass of the atmosphere is contained in the troposphere, along with virtually all the clouds, water vapor and precipitation. Ascent of an air particle from low level to the vicinity of the tropopause can occur in a few minutes in the most vigorous thundercloud updraughts, while in clear conditions the journey may take several days.

The term "troposphere" was first coined by Leon Teisserence de Bort, a French meteorologist. He was a pioneer in the use of balloons for taking temperature "soundings" of the atmosphere, as were Richard Assmann in Germany, Abbott Lawrence Roth in the United States and James Glaisher in the UK.

Clouds

Clouds characterize the troposphere; they are produced by a range of events that all lead to significant cooling of moist air. The cooling is significant in the sense that it is intense enough to lead to saturation, after which further cooling will bring about condensation in the form of cloud droplets, or ice particles in very cold conditions.

Lapse rates

The ascent of bubbles or layers of air leads to cooling associated with their gradual expansion – this is termed adiabatic expansion or cooling. It means that, for example, as a "bubble" expands when it rises, its internal energy – or the rate at which its molecules whizz around – decreases. This leads to a drop in its temperature. If the bubble is cloudless, it will cool at a fixed rate known as the dry adiabatic lapse rate (DALR), which is 9.8°C/km [28°F/mile].

Conversely, where air sinks toward the surface, it will be compressed and warmed at this rate. If ascending air is damp enough to produce cloud droplets, latent heat will be released, which warms the air and offsets the DALR. This reduced rate of cooling is called the saturated adiabatic lapse rate (SALR) and varies according to the quantity of water vapor contained in the air. It is the rate of temperature decrease that would be measured within a cloud.

Other processes can also affect the air's temperature, such as radiative warming and cooling. In this case, gas molecules absorb solar radiation and are warmed by it, then cool by radiating heat away in all directions.

The stratosphere

This layer is distinctly different from the constantly churning troposphere below it. Its very name indicates that it is stratified, or stably layered, being a region within which temperature is constant or increases with height. Colder below and warmer above suggest that overturning motions in the stratosphere are very much reduced in stark contrast to the troposphere (warmer below and colder above). The heating at the Earth's surface, at the base of the troposphere, may be likened to heating a pan of soup or some fluid on a cooker ring. Heat is transferred partly up through the fluid by convection – in the case of the atmosphere, this happens as thermals or as "bubbling" cumulus clouds.

Sometimes, extremely vigorous and thus very deep cumulonimbus clouds within the troposphere – in the tropics or over the interior of the USA in summer for example – can actually overshoot into the lower reaches of the stratosphere. It is so stable and so very dry in this region, however, that the upward shooting cloud is soon evaporated by mixing with the ambient air.

The stratosphere was discovered independently by two European scientists in 1902. Richard Assmann and Leon Teisserence de Bort both established that above about 10 km [6 miles] the air temperature either remains constant with height or actually increases. The term "stratosphere" was applied to this layer by de Bort.

It extends from the tropopause up to 50 km [80 miles] above sea level, where its maximum temperature is reached. As an annual average above middle latitudes, this is about 0°C [32°F].

Ozone

Increased temperature in the middle and upper stratosphere is caused by the absorption of short wave solar radiation by ozone. Indeed, the very existence of ozone at these levels is due to the splitting, or dissociation, of oxygen molecules into two oxygen atoms by the action of that same radiation.

This means that ozone is constantly being created and destroyed by natural processes in the stratosphere, mainly at

Ascent and descent of a dry air bubble at the dry adiabatic lapse rate

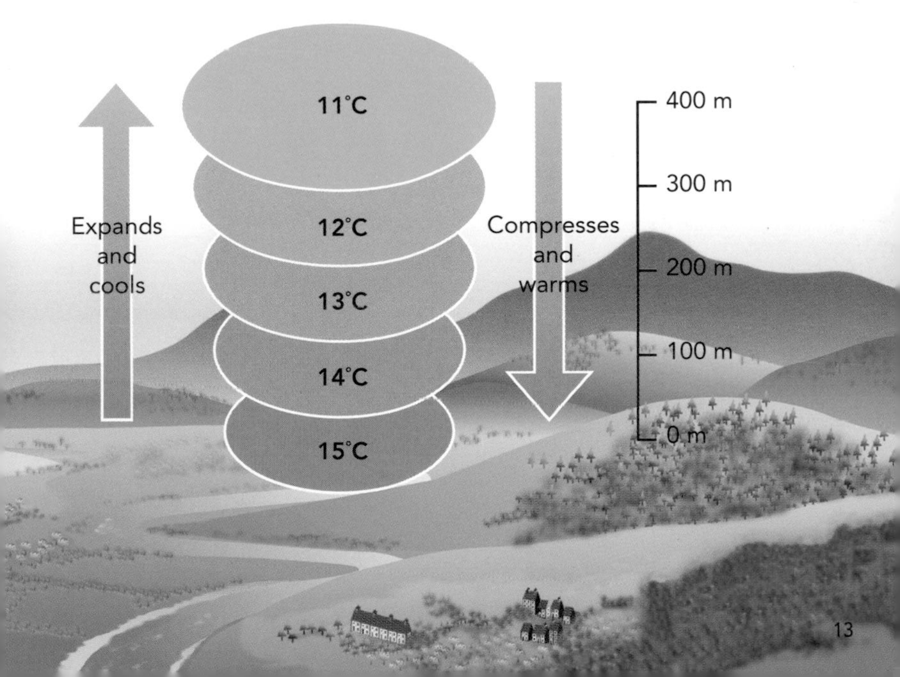

a height of 20–30 km [12–18 miles]. This has been happening for millions of years and has established a natural balance of the gas in the lower to middle stratosphere.

The complex chemical reactions that occur within the ozone layer mean that some 90% of the potentially harmful ultra-violet radiation that streams into the atmosphere in the solar beam is absorbed. Today, the artificial destruction of ozone is of enormous concern internationally, since such depletion will lead to the increased risk of harmful ultra-violet radiation reaching the Earth's surface.

Mother-of-pearl clouds

Although the stratosphere is dry, on rare occasions clouds may form within it. These are termed nacreous or "mother-of-pearl" clouds. They are believed to be made of ice crystals, and to form at the smooth crests of vertical wave motions within the lower and middle stratosphere.

Typically, nacreous clouds occur up to about 30 km [18 miles], are lenticular (lens-like) in form and exhibit a very fine, delicate structure. They are usually stationary, suggesting forcing by mountain massifs, and often display brilliant iridescence. This occurs as tinted areas that are most often red and green, but occasionally blue and green; these are caused by the diffraction of sunlight by very small cloud particles. Nacreous clouds are not significant in terms of weather, but they play a crucial role in the dramatic formation of the Antarctic ozone "hole" (Chapter 9).

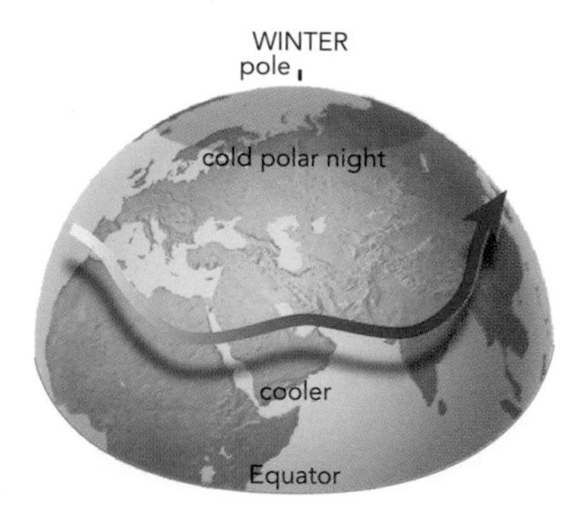

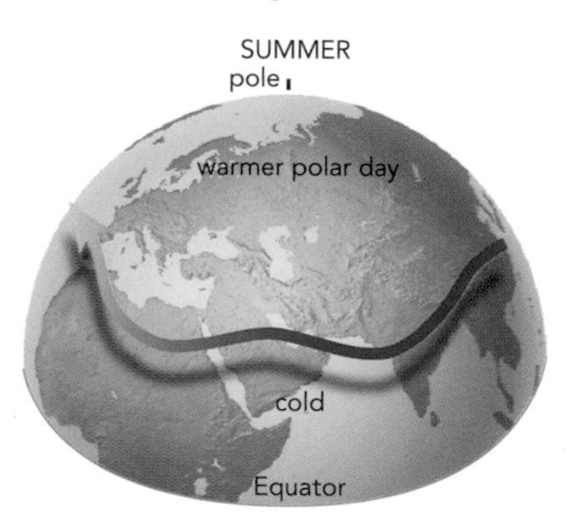

Reversal of stratospheric winds from winter (left) to summer (right)

Polar day and night

In the polar stratosphere there exists a marked seasonal change in the air temperature, which is caused mainly by the prolonged months of darkness during the polar night and the similarly extended period of light during the polar day. This means that at stratospheric levels, the polar region is colder than lower latitudes in the winter, but warmer in the summer.

Noctilucent cloud, Boroughbridge, Yorkshire, UK

This seasonal flip in the temperature gradient between the pole and lower latitudes is associated with a change in the wind circulation at these levels. Thus during the depth of the polar night, there are strong westerlies, while during the height of the polar summer, the circulation is weaker and easterly.

The mesosphere

The "meso", or middle, region lies above the stratopause and impinges on the lower ionosphere. It is characterized by temperature that decreases with increasing height, from something like 0°C [32°F] at its base to around –90°C [–130°F] at the mesopause, where the atmospheric pressure is about 1/100,000 of the sea-level value.

This profile, which is similar in pattern to the troposphere, promotes vertical circulations that occasionally lead to cloud formation over polar regions in the summertime. Typically, the cloud occurs at elevations of 80–85 km [50–55 miles] and can be viewed with the naked eye only around twilight against a dark sky, when the troposphere is mainly cloud-free. This cloud is high enough to be illuminated when the surface of the Earth is shrouded in darkness, being called noctilucent for this reason. Most commonly, it may be observed polewards of 50 degrees latitude, around the mid-night hours during summer.

The thermosphere

This deep layer stretches from the mesopause to the outer limit of the Earth's atmosphere; it lies above the well-mixed turbosphere (or homosphere), and is also termed the heterosphere. It is characterized by increasing temperature with elevation, such that at heights between 300–500 km [190–310 miles], it reaches between 500°C–2000°C [900–3600°F]. This temperature range is directly attributable to solar activity, the highest values being associated with an active Sun. It is within the thermosphere that the gases separate out according to their molecular weights (page 8).

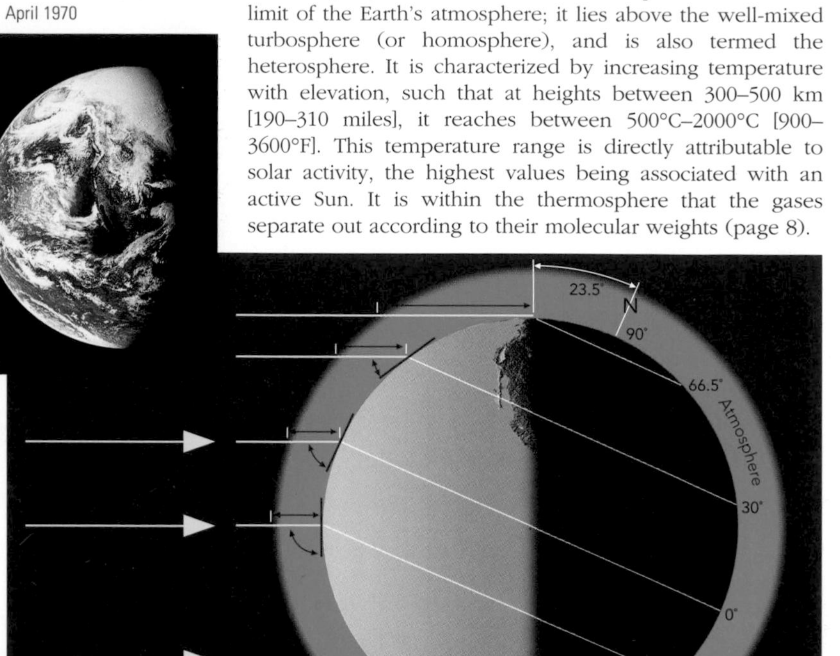

Apollo 13 photograph taken in April 1970

Varying path length and incidence angle of the solar beam in southern hemisphere summer

The Sun, radiation and the atmosphere

The virtually ceaseless motion of the Earth's atmosphere and oceans, both on global and very local scales, is ultimately related to the massive and endless stream of energy leaving our local star, the Sun. The Solar Constant, or the energy available in the solar beam at the outer limit of the atmosphere, is 1.38 kW/m^2 at right angles to the beam. This energy is transferred through space by the process of radiation. Unlike other forms of heat transfer, such as convection and conduction, this does not require a physical medium for it to be effective.

The radiative energy emitted by anything depends on its temperature. The hotter a body, the more energy it will emit, and it will do so at shorter wavelengths (the wavelength is the distance between two adjacent crests in the wave pattern that characterizes radiation). This measurement is the means by which the radiation is defined across a wide electromagnetic spectrum, from the extremely short ultra-violet to the very long waves which are used to broadcast radio programs, for example.

A body such as the Sun, with a surface temperature of about 6000K (Kelvin: 0°C [32°F] is equivalent to 273K) has an emission curve that peaks at short wavelengths. Our planet, with a mean atmospheric temperature of about 255K, emits very little in comparison and at longer wavelengths. However, this terrestrial radiation is crucially important in the heat budget of the planet's surface and atmosphere.

The electromagnetic spectrum with "classes" of radiation

In an average year, the amount of solar energy absorbed by the Earth's surface and atmosphere varies markedly from equator to pole. There is little variation within the tropics, where about 300W/m^2 flows in, while in polar areas, it falls to below 100W/m^2. This pattern is dictated partly by the angle of incidence of the incoming sunshine (solar radiation) and the thickness of the atmosphere through which it must travel. Low sun elevation and long atmospheric path length for the sunshine combine in higher latitudes to produce small values. In contrast to this steep solar curve, the Earth and its atmosphere emit their terrestrial radiation to space so that, although more is being emitted in the warmest, tropical, zone, the energy falls off gently toward the poles. In this case, emission ranges from about 260 to 120/180 W/m^2.

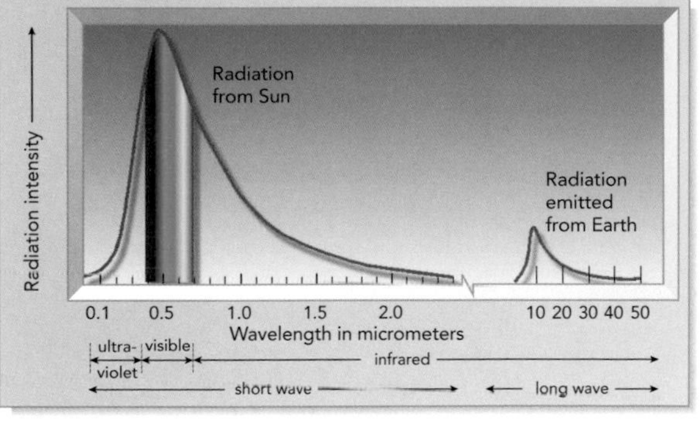

Solar and terrestrial radiation emission spectra

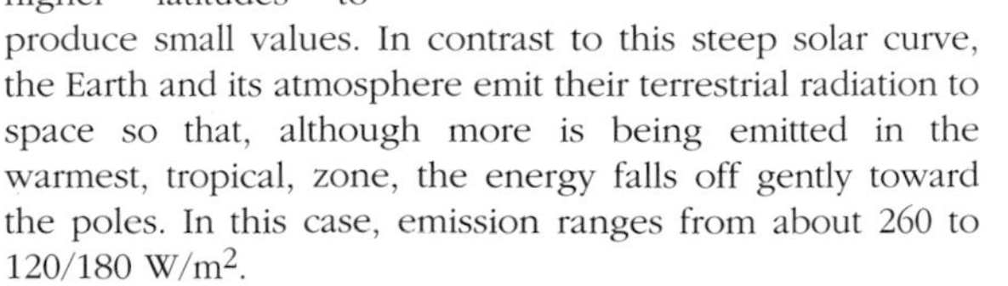

Surplus and deficit

The two illustrated curves point to the presence of a region of the Earth that, on this annual average basis, receives more solar radiation input than it loses in terrestrial output. This surplus occurs between about 35 degrees north and 35 degrees south. Conversely, the regions in each hemisphere outside these subtropical latitudes experience a radiative deficit where more energy is lost to space than is gained from the Sun.

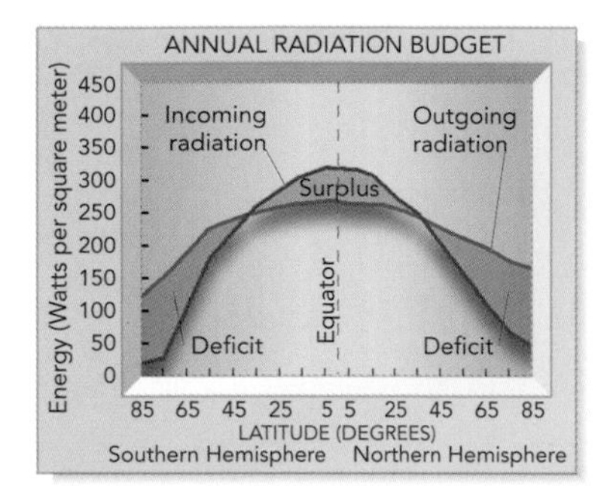

Annual average incoming solar and outgoing terrestrial radiation

These patterns pose a problem. Why does it not become progressively hotter in the tropics as the years go by, and colder in the "extratropics"? Well, if there were no atmosphere or oceans, and if the Earth were simply a ball of rock, then the tropics would become progressively hotter, to such an extent that the outgoing Earth radiation would balance the incoming flow from the Sun. Similarly, the extratropics would become colder and colder, until the Earth's outgoing energy fell sufficiently to balance the supply from the Sun.

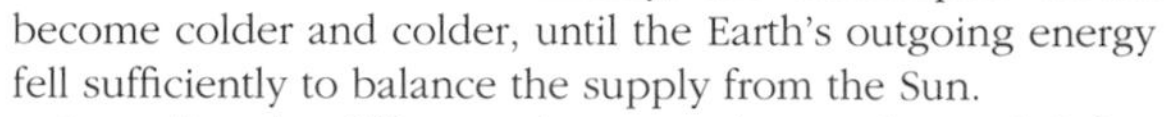

In reality, the difference between the surplus and deficit at low and high latitudes forces the atmosphere and ocean into carrying some of the excess heat from the tropics to the deficit zone outside, acting like massive convectors. This word is appropriate because the heat is transported

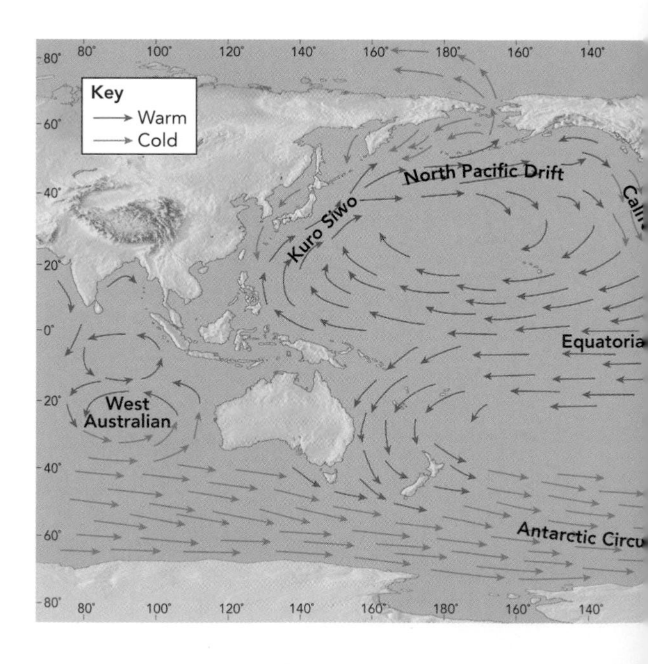

by the motions within the body of the fluid, in the same manner that a convector heater warms a room by heating the air (fluid), which then circulates around the room as a warm current. Both the atmosphere and the ocean act as global fluids, constantly ameliorating conditions in both regions – they keep the tropics cooler than they otherwise would be, and warm the extratropics above a significantly colder possibility.

The annual pattern masks large seasonal variation. Although the amount of solar radiation absorbed in the tropics varies little throughout the year, being mirrored by the limited seasonality of tropical temperature, the values at higher latitudes vary enormously. The picture for December illustrates the fact that most of the northern, winter hemisphere is in deficit, while the summer hemisphere is largely a region of surplus. The principal reason for this is the change from the prolonged polar night and "dark" mid-latitudes in the winter to the polar day and "light" mid-latitudes in the summer.

Therefore, the forcing pattern is strongest in the winter, because the difference between the equator and pole is at its greatest; in the summer, it is much weaker.

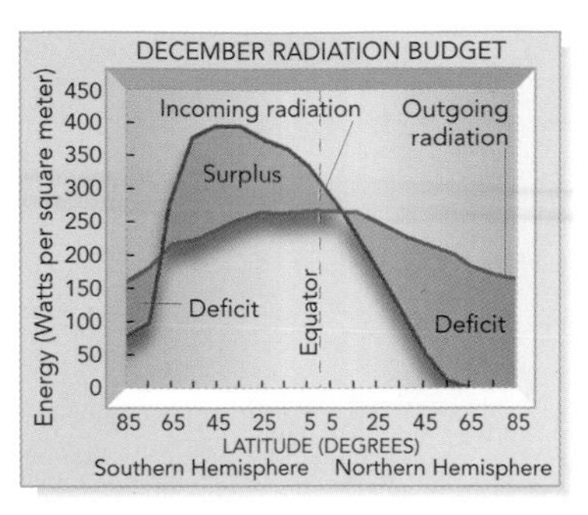

December average incoming solar and outgoing terrestrial radiation

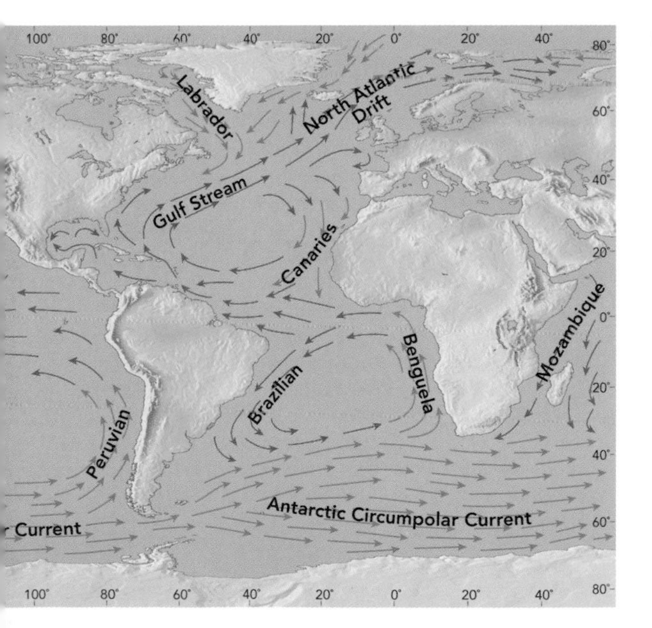

Major surface ocean currents

Working depressions

Broadly, this is the reason why the frontal depressions that run across western Europe so frequently in the cooler season are much windier (with deep pressure centers) than their warmer-season counterparts. Their increased vigor is a sign of the larger amount of heat transport that they have to undertake.

The frontal depressions that are so commonplace across mid-latitude western shores are hallmarks of the weather in those areas. Their often extensive warm-sectors (see Chapter 5) are areas of warm, moist subtropical air that are moving toward the pole and, therefore, are cooling, while behind the cold front, areas of cooler, showery polar air sweep towards the equator to be warmed. These traveling lows are nature's way of responding to the atmosphere's demand to be cooler in low latitudes, and warmer in high latitudes.

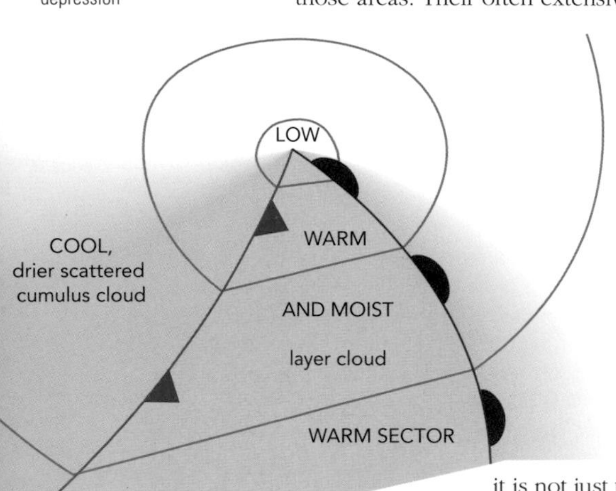

Basic contrasts in an open wave mid-latitude frontal depression

Local weather is affected by many other factors, too; it is not just the lows, or depressions, that do the work. Highs, or anticyclones, are also important, as is the circulation of the air in great depth.

Working oceans

Ocean currents play an important role in influencing the atmosphere, not only globally, but also on regional and local scales. The Gulf Stream, the North Atlantic Drift, and the Kuro Siwa are all examples of warm water being exported from the tropical "boilerhouse", and all have a significant impact on the climates of distant shores. In contrast, the cool Canaries and California Currents are parts of the grand design to transport cooler water toward the equator for warming. On their way, they influence the weather along adjacent coastlines dramatically.

The distance of a location from the sea, its elevation above sea level, its exposure and its latitude all have roles to play in the nature of the weather experienced. These factors will be explored in later chapters to help give the reader a broad understanding of the complexities of the weather and what makes it "tick".

chapter two

2

Observing the Weather

Many people take more than just a conversational interest in the atmosphere by recording weather measurements at home. In some countries, there are networks of enthusiasts who produce regular newsletters that often include members' observations. The advent of the internet has facilitated even wider links between observers on an intercontinental scale.

Measuring the weather is carried out in a variety of ways, which together form the lifeblood of the modern science. Operational meteorology relies upon a wide range of observations for the analysis of the current global weather situation and the prediction of its evolution over hours and days. To forecast, it is essential that the current state of the atmosphere is known, not only at the surface, but also throughout its depth. Observations around the world must be taken simultaneously, using the Universal Time Co-ordinated (UTC or GMT) clock, so that weather centers have a "snapshot" or synoptic view of the global, regional, and local weather.

The measurements that are taken routinely at weather stations across the globe are a standard set of observations laid down by international agreement through the agency of the United Nations' World Meteorological Organization (WMO) in Geneva, Switzerland. The greatest frequency with which the overall surface weather is observed operationally is hourly at busy airports and military airfields. At many other sites, however, observations are taken every three, six or perhaps 12 hours. In these cases, it is important that they include the hours of 0000 and 1200 UTC, because these are the key times on which many forecasts are based.

There is a significant trend in many weather services to automate all, or many, of the tasks traditionally carried out by members of staff. Doubtless, this will continue, leading to a decline in the availability of, for example, information on visibility and cloud type and the amount of data that are currently impractical and/or too expensive to assess automatically.

Normally, the surface observations reported every hour are:

- dry bulb temperature
- dewpoint temperature
- mean-sea-level barometric pressure
- pressure tendency
- total cloud amount
- cloud type and base height
- horizontal visibility
- wind direction and speed
- present and past weather
- precipitation total (usually 12- or 24-hour)

Over the last 50 years or so, global upper-air observations have developed into an essential component of the network, providing the all-important information on how temperature, humidity, wind direction, and speed vary up to about 20 km [12 miles] above sea level. These variables are monitored by balloon-borne instrument packages called radiosondes, which are released routinely four times a day – at best – at 0000, 0600, 1200, and 1800 UTC. Wind data are available at each of these hours, whereas temperature and humidity are usually recorded only at 0000 and 1200 UTC. Although there are fewer upper-air stations than surface sites worldwide, the data are crucially important to the individual forecaster and to computer-based global weather prediction schemes.

Only a few ocean weather ships remain, being dedicated to measuring surface conditions every hour and releasing radiosondes four times a day. They steam around fixed locations in the central and eastern North Atlantic, but are being substituted nowadays by roving commercial ships. Such vessels release radiosondes automatically at the appropriate time, wherever they are.

Measuring the atmosphere is not confined to surface and upper-air stations. Since 1960, weather satellites have orbited the Earth, not only providing operational meteorologists with details of the location of clouds across the globe, but also having many other useful applications. These include vertical profiling of temperature and humidity down through the atmosphere to help fill large gaps in the radiosonde network.

Radar has evolved since World War II into a useful tool for weather analysis because it can be "tuned" to sense precipitation within about 100 km [60 miles] of the antenna. Today, the United Kingdom is covered by a network of such radars, from which a national map of the extent and intensity of precipitation is produced every 15 minutes. Many other weather services have similar systems.

Temperature

The standard method of measuring air, or dry bulb, temperature is within a properly exposed weather screen, which will house other instruments too. The screen is designed to ensure that the air temperature measured really is just that – the temperature of the air flowing through it via the gaps between its sides' downward-angled slats. The reflective quality of the box's white finish combines with its insulated floor and roof to minimize any effect on the air temperature by sunshine or the temperature of the ground below the screen.

Horizontally-mounted maximum (top) and minimum thermometers, thermograph (left), and hygrograph (right)

Dry bulb and wet bulb temperature

Air temperature is measured with a mercury-in-glass thermometer that is read to the nearest 0.1°C [32.2°F]. This is known as a dry bulb thermometer.

A second such thermometer is also housed vertically in the screen, but its bulb is snugly covered by a muslin bag that is kept permanently wet with distilled water supplied by a wick. This instrument is termed the wet bulb thermometer, and although it senses the wet bulb temperature in degrees Celsius or Fahrenheit, in fact it is a measure of humidity (see page 30).

Maximum and minimum temperature

dry bulb (left) and wet bulb (right) thermometers in a weather screen

Also housed within the screen are the horizontally-mounted maximum and minimum thermometers which are designed specifically to record the highest and lowest temperatures that occur during a specified period, often 24 hours from 0900 local time. Commonly, the maximum occurs during the mid-afternoon, and the minimum during the hours of darkness. Occasionally, however, the timing is very different, particularly outside the tropics, when airmass changes occur in association with the passage of frontal depressions.

The maximum thermometer is a mercury-in-glass instrument with a constriction in the narrow central thread, near the bulb. The mercury can expand unimpeded from the bulb and through the constriction as the air temperature increases. Once the temperature falls, however, the length of the mercury thread is preserved because the fluid is prevented from returning to the bulb by the constriction. Thus the maximum temperature is recorded and will remain so until the instrument is reset in the same manner as a clinical thermometer, which records body temperature using the same principle.

The minimum thermometer contains alcohol rather than mercury because its lower freezing point (–114.4°C compared to –38.9°C) makes it more useful in very cold regions. Alcohol expands along the thin bore of the instrument as the air temperature increases, and retreats when it cools. Suspended within the alcohol is a very thin "index", or marker, which is dragged back along the bore by the alcohol's meniscus as the air temperature falls. As the air warms once more, the alcohol will expand along the bore, leaving the index behind, and the tip furthest from the bulb will mark the minimum temperature precisely. Usually, this type of thermometer is reset once a day by gently tilting it so that the index drifts back to the meniscus.

A forecast of "tonight's low" or "today's high" refers to the minimum in a weather screen. In addition to this measurement, daily mimima are also recorded over grass, bare soil, and concrete surfaces.

Thermograph

All the measurements described so far are "point" values. In contrast, the thermograph, an old-fashioned, but still widely used instrument, provides a continuous trace of temperature, typically over a period of a week. Essentially, it consists of a pen on the end of a long arm attached to a bimetallic coil, which distorts as the air temperature rises and falls. The pen traces such fluctuations on a thermogram – a paper strip chart – wrapped around a clockwork-driven drum.

More modern electrical instruments are also used, by necessity, in automatic weather stations. However, many instruments routinely in use today were developed many decades ago.

Pressure

To measure atmospheric pressure is to weigh the great mass of air that presses down upon the Earth. The pressure decreases with height through the atmosphere, because there is progressively less air above a given level.

This downward pressure can support a column of water, or other fluid, in a glass tube immersed in a reservoir at its lower open end, and topped by a vacuum at its upper sealed end. Atmospheric pressure is such that at sea level, this water column would be about 10 m [30 ft] high – rather impractical for a weather station! The high density of mercury means that its column height is a much more manageable 75 cm [30 inches] or so.

Mercury barometer

The mercury barometer was developed during the 1640s by Evangelista Torricelli, a student of Galileo. It is still widely used, but its reading must be corrected for the influence of the surrounding air temperature and local acceleration due to gravity. Both affect the height of the mercury column.

The "station level" pressure is read to the nearest 0.1 mbar (0.1 'hectopascal' or hPa), but this reading must be adjusted to a common datum, which is mean-sea-level. This entails adding a certain number of millibars to represent the pressure of an imaginary column of air between the barometer and mean-sea-level. If this was not done, the resulting weather chart would look like a topographical map. This is because the rapid change of pressure going up through the atmosphere (about 1 mbar every 10 m [30 ft] near sea level) would completely conceal the much more subtle change of pressure across the surface caused by genuine highs and lows (typically 1 mbar every 100 km [60 miles]).

If a weather station is below sea level the adjustment is made by subtracting so many millibars from the reading.

Mercury barometer

Aneroid barometer

Many homes have what are known as aneroid (without air) barometers. This type senses the pressure through small distortions of a partly evacuated metal capsule. Higher atmospheric pressure will "squash" it more than lower pressure. The capsule is linked mechanically to the familiar arrow that moves around a scale of millimeters and/or inches of mercury, and of millibars. In addition, the over-simplified, and often unreliable, forecasts of "Dry", "Change", "Wet", etc. are printed on the face of the instrument alongside the scales.

Barograph

The drawback of a barometer is that it provides only an indication of pressure at the time it is read, which is of limited use. More significant is the rate of change of pressure with time at each station and the pattern of pressure across the surface at mean-sea-level, because weather-producing features such as highs and lows are identified by its routine mapping.

Barograph with a week's trace of pressure change

The change of pressure with time is portrayed by a barograph, an aneroid instrument with an indicating arm that traces a continuous line of pressure on a barogram – a paper strip wrapped around a clockwork-driven drum. The barogram is usually changed once a week.

In addition to the "spot" value of pressure, the barograph also provides its "tendency", because it is very useful to know the size and direction (up or down) of a station's pressure change. Typically, this covers the three hours leading up to the observation time.

Precipitation

This term encompasses all forms of water particle that fall from the atmosphere to the Earth's surface. Apart from rain, it includes drizzle, snow, and hail.

Graduated measuring vessel (right) and standard 5″ raingage (left)

Daily gage

The most common instrument for measuring precipitation is a raingage that is emptied once a day to provide a simple record of fall in millimeters or inches. The design varies little from country to country: often it comprises a 12.7 cm [5 inch] diameter copper cylinder with its top 30.5 cm [12 inches] above the surrounding surface. This height reduces the risk of water splashing in from the ground and aids the retention of snow.

Precipitation that falls into the gauge runs down a funnel with a narrow aperture to minimize evaporative losses. It is collected in a vessel (often a glass bottle) that is sunk into the ground. Once a day, the observer decants the water into a tapered glass measuring vessel to determine the amount to the nearest 0.1 mm or 0.01 inch.

Gages should be sited well away from any objects, such as buildings, that may affect the natural trajectory of any precipitation. In a sense, all are "active" instruments because their presence affects airflow and thus the drift of precipitation, which would fall differently if the gages were not present.

Recording gage

Daily totals are useful, but do not provide information about the intensity and duration of precipitation, which is essential for detailed studies. Autographic or recording raingauges are designed to satisfy the requirement for such detail on a strip chart that is usually changed daily, or automatically as a telemetred radio message from the gauge to a central point. A common type is the "tipping bucket" design, in which two small open metal containers on a see-saw mechanism are used to collect the precipitation. When one bucket is filled by the required amount, it tips, moving the other into position to continue collecting any further precipitation. The tipping action is registered on the strip chart trace.

UK and Irish precipitation radar sites. Inner circle: best quality data; outer circle: limit of radar view

Radar

Precipitation totals can vary significantly over very short distances, so the spot values recorded by raingauges cannot give a complete picture. The solution lies in the images, or maps, provided by precipitation radars. These make a circular scan at a variety of shallow inclined angles every 15 minutes, or less frequently. They emit pulses of

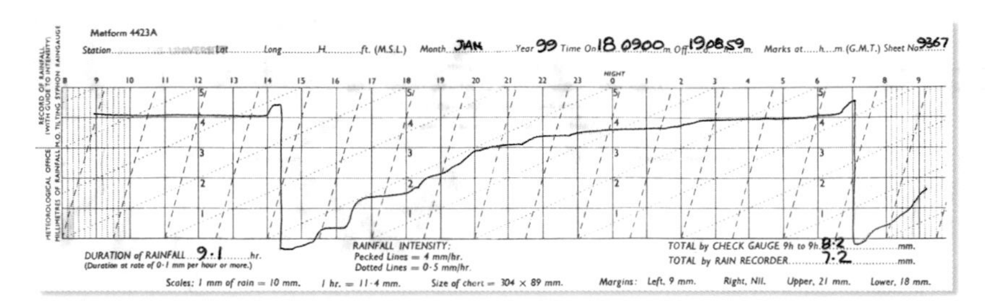

Autographic daily raingage strip chart

radiation, small fractions of which will be reflected back to the antenna by precipitation-size particles (not by cloud droplets, which are smaller). The radar converts the ref lected radiation into an image showing rainfall rate. Radars display rainfall out to a radius of some 150 km as a pattern of boxes 1 km square, each of which contains an instantaneous average rainfall rate.

The effectiveness of radar mapping depends on the type of precipitation and its distance from the radar. The nature of the radar itself and the prevailing atmospheric conditions also have an effect.

While gauges register precipitation at the surface, radars can only measure the extent and intensity of precipitation on its way down. This is because a radar's field of view is affected by "ground clutter" (buildings, hills, etc.). Consequently, the raw data produced by radar must be modified to represent surface values by using a small number of "check" gages.

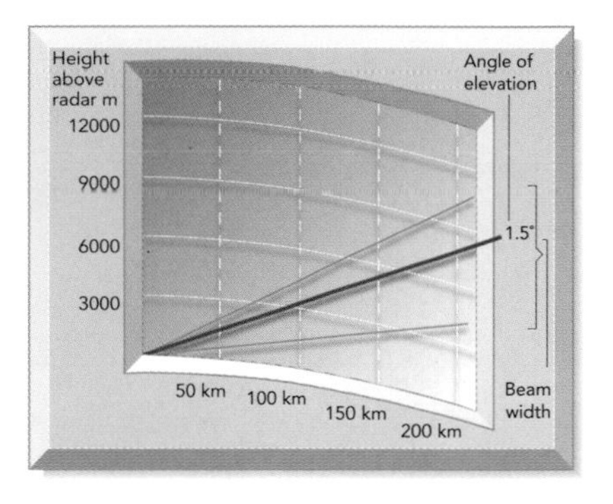

Spread of radar beam (2° width) with 1.5° elevation as a function of height and horizontal distance

Because the radar beam is emitted at a shallow inclined angle, it views progressively higher elevations the further it is from the radar. This means that it may miss precipitation in some areas.

Weather services in developed countries have either complete geographical coverage by precipitation radars, or regional facilities that provide mapping only in heavily populated areas.

Humidity

The measurement of humidity that springs to mind for most people is that of relative humidity. In fact, this is one of its least useful definitions.

Some of the definitions of humidity used in meteorology are listed below:

Absolute humidity The maximum amount of water vapour (in grams) that can be contained in a cubic meter of the air and water vapor mixture.

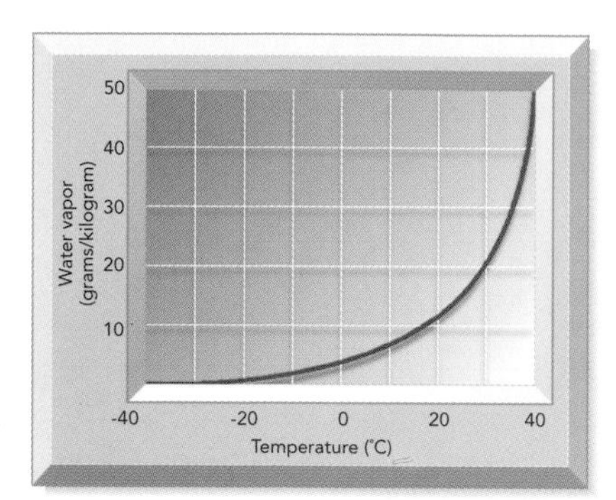

Saturation specific humidity as a function of temperature

Specific humidity The mass of water vapor (in grams) in a kilogram of the air and water vapor mixture.

Mixing ratio The mass of water vapor (in grams) present in a kilogram of dry air.

Vapor pressure The pressure exerted at the Earth's surface by water vapor contained in the atmospheric column. This varies from virtually zero to 3% of the total pressure, which is typically 1000 mbar.

Relative humidity The ratio, expressed as a percentage, of the actual amount of water vapor contained in a sample of air to the amount it could contain if saturated at the observed dry bulb temperature.

Absolute measures

The wet bulb temperature forms the basis of the calculations for relative humidity and absolute humidity. The specific humidity of air that is saturated with water vapor increases with temperature. Saturated air with a temperature of 0°C [32°F] contains 3.0 g/kg; at 10°C [50°F] this rises to 7 g/kg; at 20°C [68°F] it is 14 g/kg; and at 30°C [85°F] it is 26 g/kg.

Relative humidity

Let us assume that on a summer's day we record a dry bulb temperature of 30°C [85°F] and a wet bulb value of 15°C [60°F]. The actual absolute humidity of this sample of air is deduced from the wet bulb temperature, using standard hygrometric tables. These will also specify the value of absolute humidity if the air was saturated at its dry bulb temperature (30°C [85°F]).

In this case, the observed absolute humidity is 11 g/m^3, while the saturation (maximum possible) absolute humidity is 30 g/m^3. Therefore, the relative humidity is 11 g/m^3 ÷ 30 g/m^3 = 37%.

The same value of relative humidity can be obtained from very different samples of air scattered around the globe, so it is not of great use to meteorologists. They find absolute values of water vapor much more useful because these permit significant calculations to be made – for example, how much precipitation might fall from a column of air if all or some of the water vapor in it condenses.

Hygrometer

In addition to the spot values of relative humidity provided by the dry and wet bulb thermometers in the screen, a time trace can also be provided by a recording hygrometer. These vary, but one type commonly in use is the hair hygrometer, which takes advantage of the fact that horse (and human) hair lengthens and shortens as relative humidity varies. Human hair shrinks in length by some 2.5% when relative humidity reduces from 100% to 0%.

A small sheaf of hair is stretched across a thin metal bar, which is connected mechanically to a pen that traces fluctuations in relative humidity on a hygrogram strip chart. This is wrapped around a rotating drum and normally is changed once a week.

Other hygrometers are based on the moisture-absorbing properties of various chemicals, which become moister as the humidity increases.

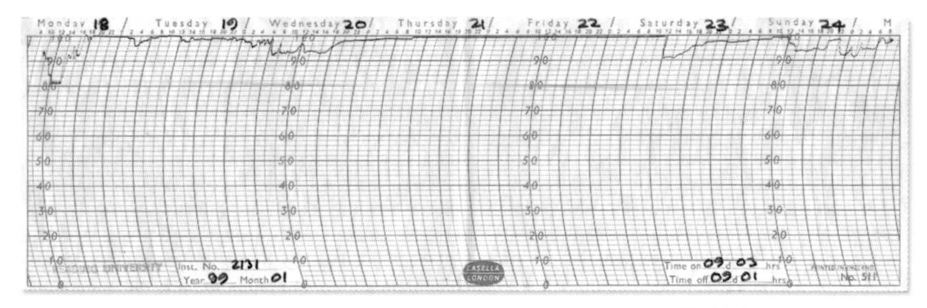

Weekly hygrogram

Cloud amount and type

By international agreement, cloud amount is reported as eighths (or oktas) of the sky covered, as both individual layers of cloud and as total cloud amount, which summarizes one or more types of cloud that may be present.

Clouds belong to one of three layers, defined simply as low, middle, or high. The layer in which they occur depends on the height of their base above the surface: the definition of these three layers varies with latitude, such that the height limits generally increase with decreasing latitude (see table on page 8).

There is a very large variety of cloud types, which are separated into four primary groups. Each of the groups contains two or three principal cloud forms which are shown on pages 34–35:

High clouds	Cirrus, cirrocumulus, and cirrostratus.
Middle clouds	Altocumulus and altostratus
Low clouds	Nimbostratus, stratus, and stratocumulus
Vertically developed clouds	Cumulus and cumulonimbus

Cumulus clouds have a "bubbly" appearance; cirrus clouds are wispy; stratus is sheet-like; and nimbus clouds are rain-bearing. These basic types can be combined, hence cirrocumulus and nimbostratus, etc. The prefixes "alto" and "cirro" are applied specifically to middle and high clouds respectively.

Anemometer and wind vane

Clouds are produced when moist air is cooled to a critical temperature, at which point further cooling leads to cloud droplets that condense from the invisible water vapor present.

There are two such critical temperatures: the dewpoint temperature and the wet bulb temperature. The former is the temperature to which the air must cool (at constant pressure with no change in the absolute humidity) for it to reach saturation point. This is not measured, but it may be deduced from the dry and wet bulb temperatures.

The wet bulb temperature is the value to which the dry bulb temperature will cool when water is evaporated into the air until it is saturated. The process of evaporation extracts heat from the air, which means that it cools.

Wind speed and direction

Anemometers

Wind speed is measured by an anemometer. The type used routinely at weather stations is the cup anemometer, which usually has three hemispherical cups mounted on a vertical shaft. The pressure exerted by the wind on the concave inner faces of the cups is greater than that on their convex outer faces, which causes the vertical shaft to rotate. The rotation rate varies with the wind speed, which is displayed on a calibrated dial marked with knots (nautical miles per hour), meters per second and other units.

A properly exposed anemometer will be mounted 10 m [30 ft] above the surface.

Horizontal and vertical wind speed

In operational meteorology, the wind-speed measurements are averaged over the course of a few minutes. The value measured is the horizontal wind speed, and while the vertical component is important, it is not measured routinely – typically, it is about 100 times smaller than the horizontal wind. In some circumstances, however, it can actually outweigh the horizontal wind speed, as with the strong vertical currents associated with very deep cumulus clouds, for example.

Beaufort scale

In 1806, Francis Beaufort, a Royal Naval officer, developed a scale to express the effect of different wind speeds on the sea's surface. Later, information was added to facilitate the scale's use over land, and by the early 20th century, numerical values of wind speed had been linked to the Beaufort numbers 0 to 12.

Beaufort scale of wind force

Force	Specifications for use on land	Equivalent mean wind speed 10 m above ground	
0	Calm; smoke rises vertically	0kt	0 $m\ s^{-1}$
1	Light air; wind direction shown by smoke drift, not by vanes	2	0.8
2	Light breeze; wind felt on face; leaves rustle; vanes move	5	2.4
3	Gentle breeze; leaves and small twigs moving; light flags lift	9	4.3
4	Moderate breeze; dust and loose paper lift; small branches move	13	6.7
5	Fresh breeze; small leafy trees sway; crested wavelets on lakes	19	9.3
6	Strong breeze; large branches sway; telegraph wires whistle; umbrellas difficult to use	24	12.3
7	Near gale; whole trees move; inconvenient to walk against	30	15.5
8	Gale; small twigs break off; impedes all walking	37	18.9
9	Strong gale; slight structural damage	44	22.6
10	Storm; seldom experienced on land; considerable structural damage; trees uprooted	52	26.4
11	Violent storm; rarely experienced; widespread damage	60	30.5
12	Hurricane; at sea, visibility is badly affected by driving foam and spray; sea surface completely white	>64	>32.7

HIGH

Tropics 6000 to 18,000 m

Cirrus

Middle latitudes 5000 to 13,000 m

MIDDLE

Tropics 2000 to 8000 m

Altostratus

Middle latitudes 2000 to 7000 m

LOW

Tropics 0 to 2000 m

Cumulus

Middle latitudes 0 to 2000 m

Cirrocumulus and cirrus

High latitudes 3000 to 8000 m **Cirrostratus** (above cumulus)

HIGH

Altocumulus

High latitudes 2000 to 4000 m **Nimbostratus**

MIDDLE

Stratus

High latitudes 0 to 2000 m **Stratocumulus**

LOW

Solarimeter

Campbell-Stokes sunshine recorder

Wind direction

Combined with the anemometer is a vane that points into the wind to show the direction from which it is blowing. Commonly, this consists of a horizontal arm with a pointer at one end and a streamlined vertical plate at the other. Movements of the vane are transmitted to an anemograph, which provides a continuous trace of direction fluctuations.

A wind direction report is usually given as an average taken over a few minutes, and it is expressed in degrees read clockwise from true north to the nearest five degrees. The value of 000° is reserved for calm conditions when there is no wind. An easterly (that is, a wind blowing from the east) has a direction of 090°, a southerly blows from 180°, a westerly from 270°, and a northerly from 360°. There are finer gradations, such as a south-westerly being 225°.

The convention in meteorology is to report the direction from which the air flows because it is important to know its past trajectory. The temperature and humidity of the air are

Haboob, Sudan

partly determined by the nature of the surfaces over which it approaches an observation site.

Sunshine

The routine method of measuring the duration of "bright sunshine" is to use a sensitized card which is held in a frame wrapped around one half of a glass sphere that focuses the Sun's rays on to it. The term "bright" indicates that this type of recorder is not sensitive enough to record sunshine around sunrise and sunset. It provides a total duration of bright sunshine to the nearest tenth of an hour every day.

A more useful measurement is that of the intensity of solar radiation received at the surface, expressed in Watts per square meter. Some observation sites are equipped with solarimeters to sense this variable, the most basic type using a device known as a thermopile, which converts heat into electrical energy.

The solar radiation takes two forms: direct and diffuse. The former reaches the instrument directly from the Sun, while the diffuse (or sky radiation) arrives after being scattered by gas molecules, dust, and other particles.

Visibility

Clearly, knowing the details of poor visibility is much more important than being aware of good visibility, because weather hazardous to traffic needs to be mapped in as much detail as possible. Visibility is the distance at which an object can be seen and identified by someone with normal eyesight under normal daylight conditions.

Visibility is an indication of the air's opacity, and it depends on the nature of the particles in suspension. These range from extremely small particles of smoke, dust, or water that settle out very slowly in light winds, to coarse particles that are kept in suspension only by the turbulence associated with strong winds. Duststorms, sandstorms, and some blizzards fall into this latter category. The former is characterized by haze when the particles are dry, or mist and fog when water is present.

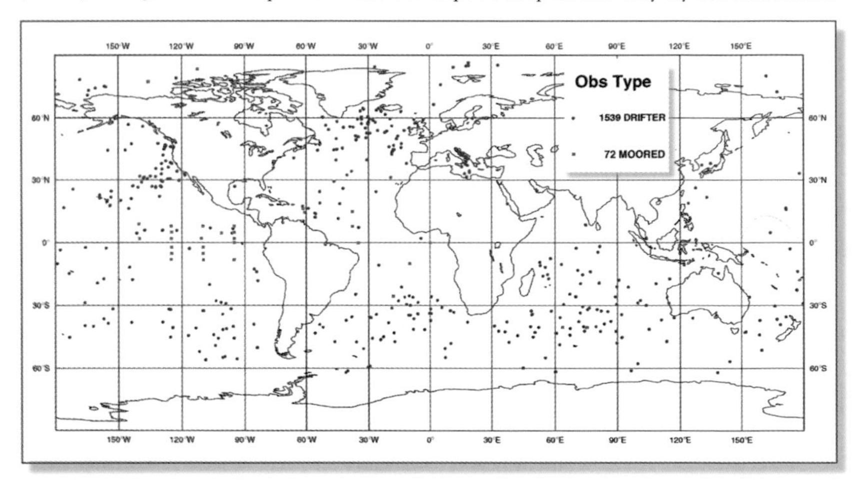

Typical distribution of drifting (red dots) and moored (blue squares) buoys for one day's 12 UTC reports received at the European Center for Medium-Range Weather Forecasts (ECMWF), Reading, UK

Over land

At synoptic stations, the observer must assess the poorest horizontal visibility (it may vary with direction from the vantage point used). At land stations, visibility is quoted to the nearest 100 m up to 5 km, then to the nearest kilometer from 5 km to 30 km, and every 5 km up to a maximum of 75 km. This is achieved by reference to objects at specific distances from the vantage point. If visibility is extremely poor – less than 100 m – it is reported to the nearest 10 m.

Radiation fog

And at sea

At sea, taking these measurements is particularly difficult because there are no objects conveniently located at fixed distances from a ship! Therefore, the scale used for marine visibility observations is much coarser, as is that for climatological stations at which values are logged only once a day.

At night, reports are often based on unfocused lights of moderate intensity at known distances and, if appropriate, the silhouettes of hills or mountains against the sky.

Commonly, these values are assessed by eye. However, there are automatic visibility meters that measure the degree of extinction of a beam measured over a short distance – typically some hundreds of meters – which is extrapolated to provide an estimate of visibility that is consistent with the routine method.

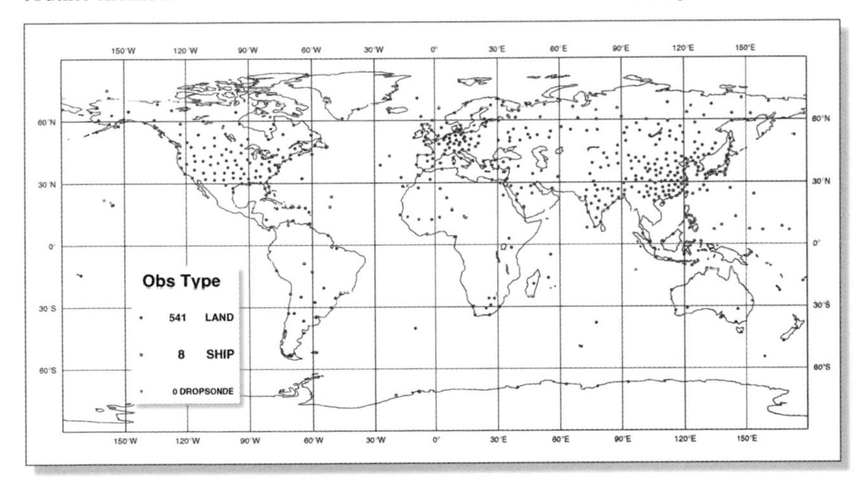

Typical distribution of 1200 UTC upper air reports received on one day at ECMWF, Reading, UK

Present and past weather

To an observer, the term "weather" does not mean simply "dry and sunny" or "wet and windy". It means something much more specific, related to prevailing conditions at the time of the observation. An observer must choose one of a hundred possibilities to report; something must be occurring, even if it is as apparently innocuous as "clouds developing during past hour". Present weather types each have a distinct two-digit number from 00 to 99 ranked so that, broadly, the more "significant" the weather, the larger the number allotted to it.

Release of a radiosonde

Past weather must also be reported. Observers may select two broad weather types from a list of ten. The period during which this factor is reported depends on the hour of observation. Those made at 0000, 0600, 1200 and 1800 UTC relate to the previous six hours, for example.

Almost all of the observations discussed in this chapter form the basis of the station plot that appears on operational surface weather maps (Chapter 3). These plots are truly international – anyone anywhere in the world can read them, if trained.

In addition to this range of observations, ships report sea-surface temperature and the speed and direction of their motion. The reason for this is that pressure tendencies reported by them are not only influenced by the motion and changing intensity of weather systems, but also by a ship's movement relative to weather disturbances.

Marine data are invaluable because there are so many gaps in weather knowledge concerning the world's oceans. The use of moored and drifting buoys has increased in recent years to provide forecast centers with improved coverage of many areas of the oceans. They house automatic sensors that report, for example, dry bulb and wet bulb temperatures, wind direction and speed, atmospheric pressure, and sea-surface temperature.

Upper-air

Today, there are 600–650 stations around the world where, twice a day at 0000 and 1200 UTC, balloon-borne instrument packages are released. These radiosondes sense pressure, dry bulb temperature, relative humidity, wind speed and

STATION P (mb)	H (dm)	LERWICK		STAVANGER		LYON		ESSEN	
Surface		250	14	325	08	350	01	350	02
925	80	260	23	315	29	205	11	360	02
850	150	260	23	315	33	185	12	060	12
800	210	260	22	310	31	190	11	305	12
700	300	280	24	310	33	200	09	295	23
600	420	285	41	310	43	215	06	295	21
500	540	280	50	310	43	230	11	295	31
400	720	290	64	310	56	250	18	295	49
300	900	305	84	310	62	260	29	290	58
250	1050	320	110	310	58	255	37	290	58
200	1200	310	59	310	56	255	46	295	60
150	1350	300	45	315	51	265	25	300	31

Upper wind observations for four upper air stations. Winds are tabulated as a function of pressure (P in mb) and height (H – average value in decameters). Winds are plotted as direction (degrees) and speed (kt).

direction as they ascend for about an hour into the lower stratosphere before the helium-filled balloon bursts. The data are transmitted to the release point as the balloon ascends and, in some cases, as the instrument package descends beneath a parachute.

Although the balloon drifts away from its release point, and the readings are taken over the course of an hour or so, meteorologists use the profiles of temperature, etc., as if they were vertical and instantaneous. The temperature and humidity data for an ascent are plotted on special diagrams, either manually or automatically, after which the forecaster can use the profiles as an aid to making a prediction.

Information on the change of wind speed and direction with height is also useful for the forecaster. In addition to the two balloons released at 0000 and 1200 UTC, wind data only are gathered at 0600 and 1800 UTC. One way of measuring the wind during a radiosonde's flight is to use radar to track a "target" hanging from it.

All upper-air observations, like surface data, are transmitted to weather forecast centers around the world so that they can be incorporated into computer-based weather prediction models. As with any form of weather data, they must have been checked for quality and have been transmitted as quickly as possible. High-quality and "fresh" observations are essential for effective forecasting.

Satellites

Polar orbiters

A significant development for meteorological observation occurred on 1 April 1960, when the American TIROS-1

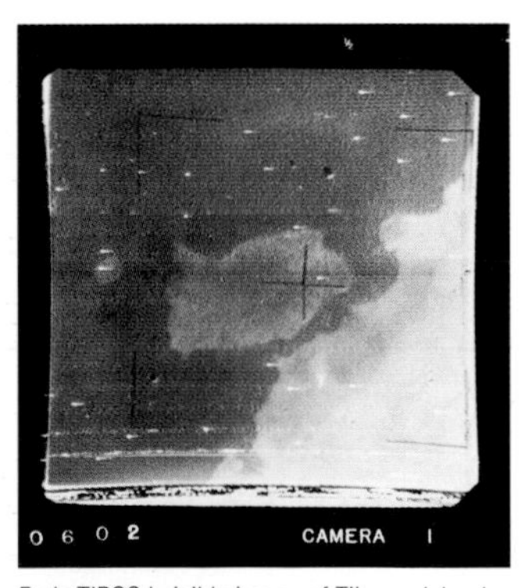

Early TIROS I visible image of Tiburon Island, Gulf of California on 12 May 1960

Thermal infrared image of a showery polar air trough west of Ireland

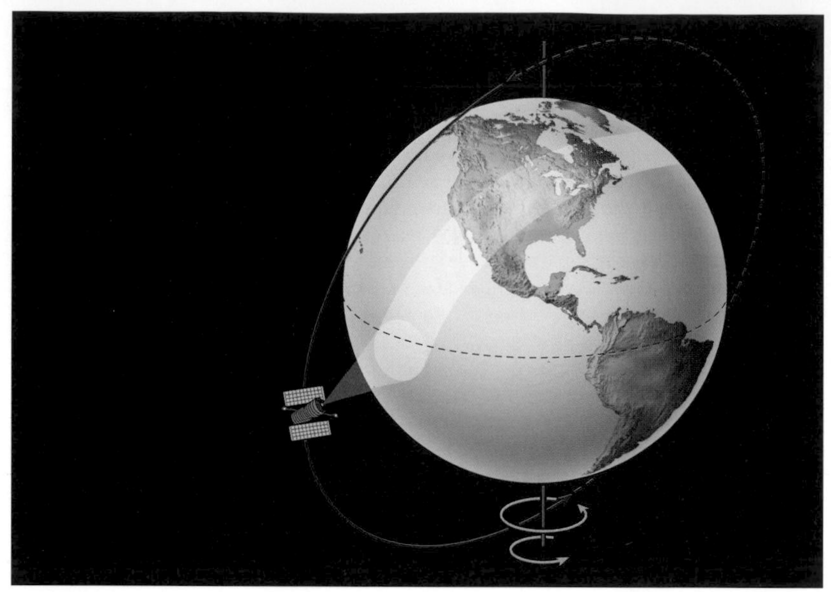

Track of a NOAA polar orbiter

satellite was launched from Cape Canaveral, Florida, USA. It entered orbit at an average height of 720 km [450 miles] and provided meteorologists with a startlingly new view of the Earth in the form of about 23,000 cloud pictures. Its operations were terminated after 78 days when its batteries were exhausted by a transmitter's failure to switch off. TIROS stands for Television and Infrared Observational Satellite and is still used occasionally to describe the current US polar orbiters that observe the planet's cloud cover. Most often, however, these are known as the NOAA satellites, after the National Oceanic and Atmospheric Administration, the US government agency that operates them.

These NOAA satellites are part of a network that forms a crucial component of the global weather observing system

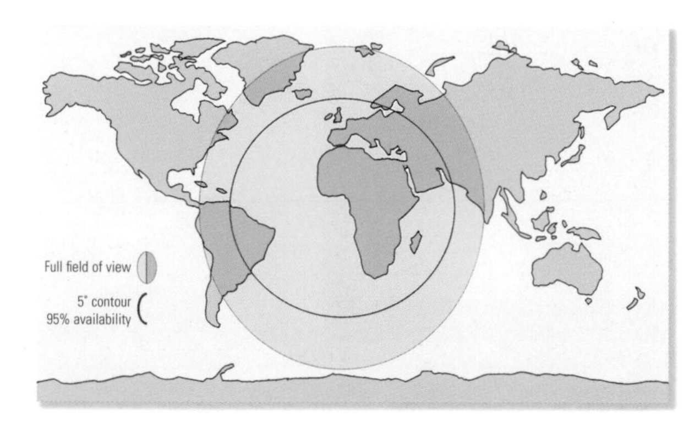

Field of view of a geosynchronous satellite

today. Currently, there are two such satellites in near-circular orbits, roughly at right angles to each other, at a height of 850 km [530 miles]. The height of a satellite determines its period – the time it takes to circle the Earth once. For a NOAA satellite, this is 102.1 minutes. Today's NOAAs weigh just over 1.7 tons and require power of 475 Watts from their solar paddles when all systems are working. There are other weather polar orbiters, including the Russian Meteor series at a higher elevation of around 1190 km [740 miles] and a period of 109.4 minutes.

Polar orbiters look down at the planet from a relatively low altitude, around 1000 km [620 miles], which is only about 0.08 of the Earth's diameter. They provide meteorologists with high-quality images along a swathe of the Earth's surface that shifts from one orbit to the next as the planet rotates beneath the satellite. For the NOAA orbiters, each swathe just touches the previous one at the equator, but overlaps progressively more toward the poles. This provides very good temporal coverage at high latitudes, but less frequent imagery within the tropics.

Geosynchronous orbiters

If a satellite is launched to 36,000 km [22,400 miles] above the equator, its complete orbit takes 24 hours. This, of course, is far higher than the NOAA satellites. At 2.8 Earth diameters from the Earth's surface, such a satellite is a long way out in space. This is called the geosynchronous (or geostationary) orbit, because the period of a satellite equals the time the Earth takes to rotate once about its axis. Thus, the satellite keeps pace with the spinning planet, racing along eastward at a speed of just over 3 km/sec [2 miles/sec] and appearing to hover above a fixed point on the equator.

This type of orbit ensures that the satellite always sees the same "full-disk" face of the Earth, producing a new image of either all or part of the region every 30 minutes. These are put together to produce an animation for a given period.

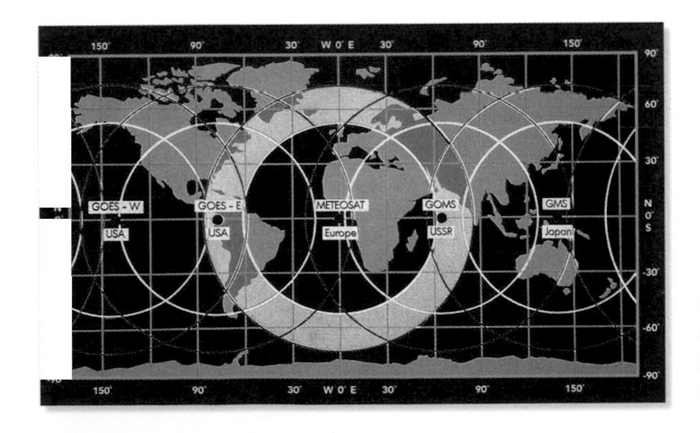

Network of geosynchronous weather satellites (GOMS now replaced by INDOEX, EUMETSAT)

There are five weather satellites distributed fairly evenly around the equator, operated by different agencies. Meteosat's are run by the European weather satellite organization, known as EUMETSAT; the two US GOES (Geostationary Operational Environmental Satellite) orbiters are overseen by NOAA and GMS (Geostationary Meteorological Satellite) is operated by the Japanese Meteorological Agency.

What do weather satellites see?

All weather satellites look down at the Earth to produce images of clouds, and many of them reveal the atmosphere's invisible water vapor, too. They do this with an instrument called a radiometer which is capable of sensing the intensity of radiation coming from the planet. The signal measured is an expression of the strength of, for example, the sunshine reflected back to space by all the surfaces being illuminated.

Visible images

As the sensor scans the Earth, during daylight hours, it will "see" very strong reflections, from fresh snow or the tops of very deep clouds. In contrast, it will sense a weak signal from cloud-free vegetated or ocean surfaces that naturally reflect considerably less "visible" solar radiation. The term "visible" is used because the waveband employed is more or less that to which our eyes are attuned. Although the images are usually processed in black and white, the scenes are what we would see (in color) if we were sitting on the satellite!

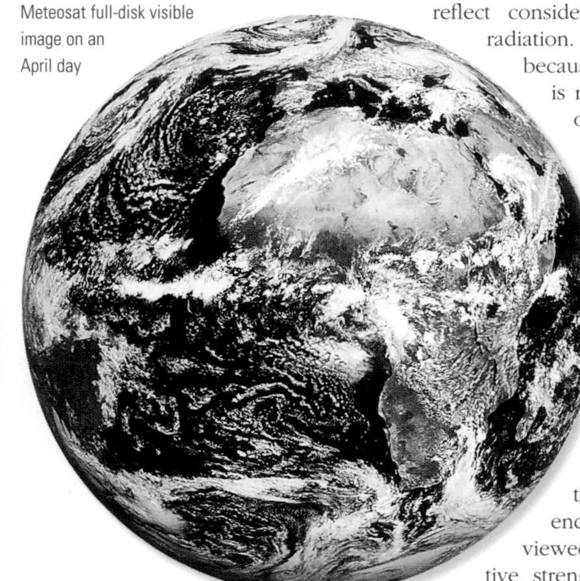

Meteosat full-disk visible image on an April day

A full-disk visible image from Meteosat illustrates cloud patterns as various shades of white. The cloud-free areas are darker shades, the tone of which depends partly on the surface viewed and its albedo or reflective strength. Generally, the ocean reflects less than 10% of sunshine falling on it and therefore appears black; the sandy Sahara is quite bright with an albedo of between 25% and 40%. Clean, dry snow reflects something like 75–95%. Clouds reflect more the deeper they are, so thin ones have albedos of 30–50%, while thicker ones are brighter with values between 60% and 90%.

One important drawback of visible images is that they are not available during the night. Weather, however, can be raging during the hours of darkness, and meteorologists need to know what is happening to the cloud-laden disturbances around the clock. This is where another common waveband comes into play.

Infrared images

In addition to monitoring short wave "visible" radiation that has come from the Sun and is reflected back into space, all weather satellites image the planet in the thermal infrared, which is a waveband that our eyes do not sense.

The strength of the infrared signal varies with the temperature of the body that emits it. This will range, for example, from an intensely hot, cloud-free Saharan surface (near 60–70°C [140–160°F]) to the frigidly cold cloud tops of equatorial cumulonimbus thunderclouds at −70°C [−90°F], or even colder! The hotter a body is, the stronger will be its signal.

Meteosat full-disk thermal infrared image on an April day

Infrared data are mapped and displayed as a black-and-white image in such a way that a strong signal appears black, and a weak one white. This way, clouds, which are colder than most other surfaces, stand out as white features.

One great advantage of the thermal infrared is that it is emitted constantly and, therefore, provides round-the-clock images. The thermal infrared image that matches the "visible" image for the same day illustrates the differences between these two wavebands.

Do they do anything else?

Weather satellites also sense additional wavebands, including one from which images of water vapor can be produced. NOAA polar orbiters monitor five channels, while the new series of Meteosats, due to be initiated in the next few years, will image the Earth in over ten.

In addition to imaging missions, some weather satellites provide thousands of vertical profiles each day of temperature and humidity down through the atmosphere. These are used in computer forecast models, as are the thousands of cloud-drift winds produced daily by the geosynchronous orbiters. Meteosat generates these automatically from sequences of three half-hourly full-disk images centered on 0000, 0600, 1200, and 1800 UTC.

At 1130 UTC, for example, specific cloud types that are known to be reasonably good "tracers" of the wind are classified automatically on the image. The shapes of these areas are automatically located on the next image 30 minutes later, and on the third image at 1230 UTC. The scheme then estimates the direction and speed of the cloud masses by comparing the first and third images – and allots a value to the middle image time. In this way, a significant number of extra wind observations are made available for regions where there are very few or no radiosondes.

Data relay

Most weather satellites provide a reliable means of communicating weather and other environmental data in real time from remote surface sites. In addition, Meteosat relays observational data, weather analysis and forecast charts around the clock to national meteorological centers in Africa, for example, where they are often not received by the more traditional landline system. Charts are "uplinked" from the UK's Met Office in Bracknell, Berkshire, while observational data are transmitted from the French and Italian HQs in Toulouse and Rome.

Automatic weather buoy

chapter three 3

Mapping the Weather

Portraying the elements of both the "actual" weather and its forecast state as a "synoptic" map, for one instant, is crucial in operational meteorology. Satellite and radar 'images' also aid the visualization of extensive detail in a quickly digestible form. Much of the complexity of the atmosphere still is, and will remain, summarized in map form that is a truly international language for meteorology.

The surface chart

The international collaboration that ensures the global transmission of weather observations is extended to the way in which weather maps are plotted. Once received, the various observations are plotted either automatically or by hand on to surface charts. No matter what the scale, however, the manner in which they are represented graphically is by international agreement. This means that charts are produced in a truly international "language".

Synoptic charts, which portray a "snapshot" of the surface weather across a region, are the working material for many forecasters. They summarize succinctly a vast amount of detail about the weather that can be appreciated very quickly indeed.

So, how do meteorologists represent all of their observations on a map? They do so by means of an agreed "station model", around which all the weather measures are plotted at specific points. Some of the values are coded while others are "actual".

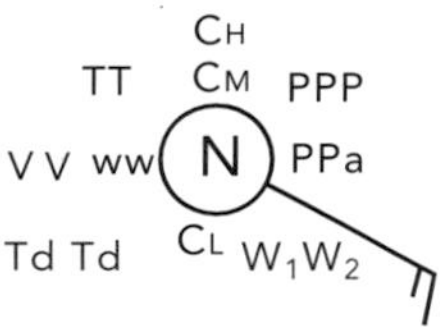

C_H
TT C_M PPP
VV ww N PPa
Td Td C_L W_1W_2

Plotting convention for surface station on land

Temperature

To the top and to the left of the station is the dry bulb temperature (TT) to the nearest whole degree Celsius, while the dewpoint temperature is plotted at bottom left. The dry bulb temperature is taken from the screen, but the dewpoint is deduced from this and the screen wet bulb temperature.

Humidity

This is represented by the dewpoint temperature (T_dT_d), which is the temperature to which the air must cool – at constant pressure and humidity mixing ratio – for it to reach saturation. Any further cooling will lead to condensation of some of the water vapor and a consequent lowering of the dewpoint temperature. The closer the dry bulb and dewpoint, the larger the relative humidity.

Using hygrometric tables, it is possible to determine the relative humidity for a given dewpoint and dry bulb temperature. In addition, for a given pressure, the humidity mixing ratio of the air can be established from its dewpoint temperature.

Pressure

Mean-sea-level pressure (PPP) is plotted to the top and right of the station. It is given to the nearest tenth of a millibar but without the decimal point and the units representing hundreds and thousands. This means, for example, that a pressure of 1035.7 mbar will be plotted as 357. Similarly, a value of 987.2 is written as 872.

There is never any uncertainty as to whether 357 means 1035.7 or 935.7 mbar; the analyst always knows which it is from the context of the chart. The main problem caused by these coded values occurs when isobars are being drawn on the maps: the meteorologist must remember that one or two digits are missing. Isobaric analyses are often drawn automatically nowadays.

Immediately to the right of the station is the pressure tendency (ppa), which represents the net change of msl pressure during the previous three hours (in mbars and tenths) along with the nature of the change. Thus, 31/ means 3.1 mbar rising, 103\ means 10.3 mbar falling, and 12 ✓ means 1.2 mbar falling then rising more.

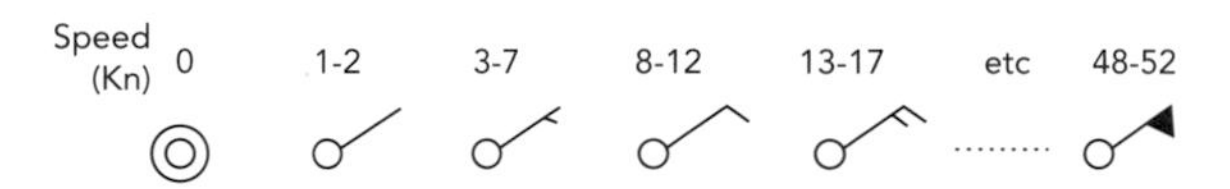

Plotting convention for wind observations

Wind direction and speed

If conditions are calm, the wind has no speed and, therefore, no direction. This is noted by a concentric circle around the station circle. Otherwise, the wind direction is represented as a "shaft" that points toward the center of the station circle along the direction from which the wind blows. A northerly wind (direction 360 degrees), for example, will be shown by a shaft that runs from due north down to the station, while a south-westerly (from 225 degrees) will be represented by a shaft that points to the center from the bottom left. Wind direction is reported to the nearest five degrees and, ideally, plotted with that precision.

The speed of the wind is reported to the nearest knot (roughly 0.5 m/sec [2 ft/sec]) and is represented as a "barb", or barbs, at the end of the direction shaft. Wind speeds are rounded to the nearest five knots before plotting. In the northern hemisphere, the speed barbs are drawn on the left side of the shaft, looking toward the station circle; in the southern hemisphere, they are drawn on the right side of the shaft.

Cloud amount

Cloud cover is assessed to the nearest eighth, or "okta", of the sky. Zero means no cloud at all, while eight eighths indicates total cover or overcast. A small bit of blue or a starry patch gives seven eighths and, believe it or not, a recording of nine eighths is valid, too! It means that the sky is obscured – by fog or a duststorm, for instance – so the observer cannot report the actual cloud cover.

Code figure	N
0	○
1	⦶
2	◔
3	◔
4	◑
5	◑
6	◕
7	◕
8	●
9	⊗
/	⊖

Total cloud amount

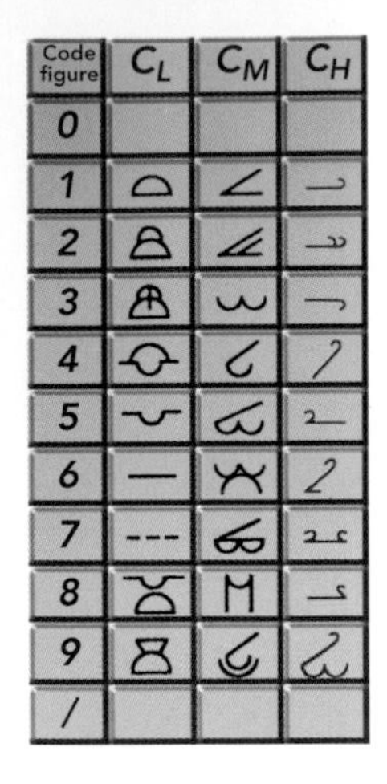

Low, middle, and high cloud types

The total amount of cloud (N) is often made up of overlapping layers and is represented by shading the station circle as shown.

Cloud type

The international code for cloud types provides 27 different symbols spread evenly between the three levels (see table on page 51).

Once all the types of cloud have been logged, they are plotted such that any low cloud (C_L) is shown below the station circle, any middle cloud (C_M) immediately above it, and any high cloud (C_H) above the middle cloud. On days when there are many different cloud types, the plotted station looks very "busy"!

Visibility

This is indicated by a two-digit code (VV) to the left of the station circle, outside the present weather symbol if there is one. Since information on poor visibility is of greater use than fine gradations of good or very good visibility, the code is organized so that half of the range of numbers relates to visibility at, or less than, 5 km.

Therefore, plotted values from 01 to 50 represent visibility in tenths of a kilometer: 41 means 4.1 km and 25 indicates 2.5 km. Numbers between 51 and 55 are not used, while from 56 to 80, the increment becomes 1 km after subtracting 50 from the plotted number! Thus, 57 means 7 km and 73 is used for 23 km. The range from 80 to 89 increases in jumps of 5 km such that 80 is 30 km, 81 is 35 km and so on. Finally, the range from 90 to 99 is reserved for visibility observations taken from ships, drilling rigs, and some coastal stations where fine resolution is not possible. The poorest visibility would be given as 90, and the best as 99.

	LI	LC	MI	MC	HI	HC
So, for snow	✻	✻✻	✻ ✻	✻ ✻✻	✻ ✻ ✻	✻ ✻✻ ✻

The abbreviations mean:

LI/LC	light	intermittent/continuous
MI/MC	moderate	intermittent/continuous
HI/HC	heavy	intermittent/continuous

There is a different symbol for showery precipitation to distinguish it from the above. So, for light showers the following pertain:

rain shower ●▽ snow shower ✻▽ hail shower △▽

Present and past weather

If there is "weather" when the observation is made, a symbol (ww) is plotted immediately to the left of the station circle. The amount of precipitation is not usually represented on a synoptic chart. However, the occurrence of precipitation – either when the observation was made or since the previous observation – is plotted. In addition, precipitation can be included as an ingredient of the past weather.

Rain, snow, or drizzle that falls from extensive sheets of cloud is indicated by the use of symbols that represent both the duration and intensity of the precipitation. These are illustrated opposite.

The table below shows all the symbols used and their broad definitions.

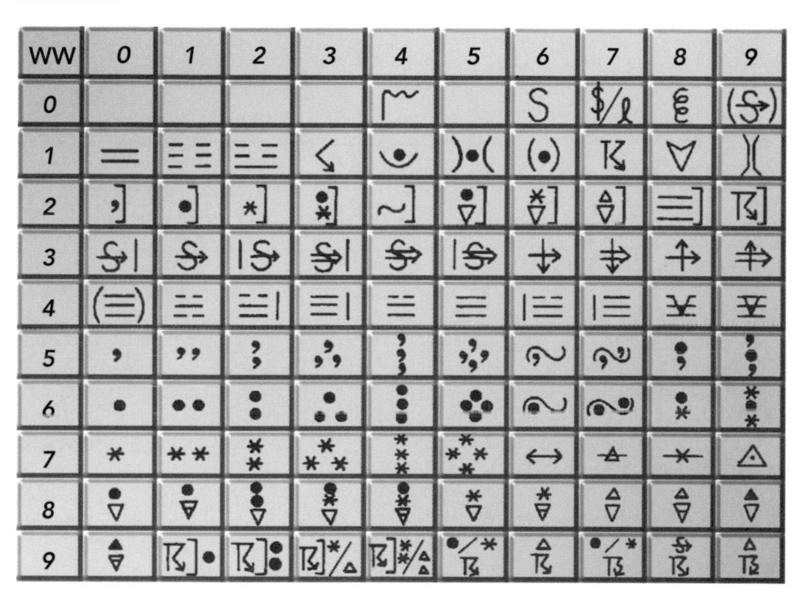

Present weather

00 to 19	No precipitation at the site at the time of observation.
20 to 29 the	Precipitation, fog, or thunderstorm at site in the past hour, but not at the time of the observation.
30 to 39	Duststorms, sandstorms, drifting or blowing snow.
40 to 49	Fog or ice fog.
50 to 59	Drizzle.
60 to 69	Rain (68 and 69 are sleet).
70 to 79	Non-showery solid precipitation.
80 to 90	Showery precipitation.
91 to 94	Precipitation, but with thunderstorm in the past hour.
95 to 99	Precipitation with thunderstorm.

Past weather (W_1W_2) is plotted to the bottom and right of the station with a choice of ten sorts:

0 Cloud covering half or less of the sky throughout the period.

1 Cloud covering more than half of the sky for part of the period, and less than this for the rest.

2 Cloud covering more than half of the sky throughout the period.

3 Sandstorm, duststorm or blowing snow.

4 Visibility less than 1 km [0.6 miles] because of fog, ice fog, or thick haze.

5 Drizzle.

6 Rain.

7 Snow or mixed rain and snow.

8 Showers.

9 Thunderstorm with or without precipitation.

Code figure	W_1W_2
0	
1	
2	
3	S→ / +→
4	≡
5	,
6	•
7	*
8	▽
9	⌈↘
/	

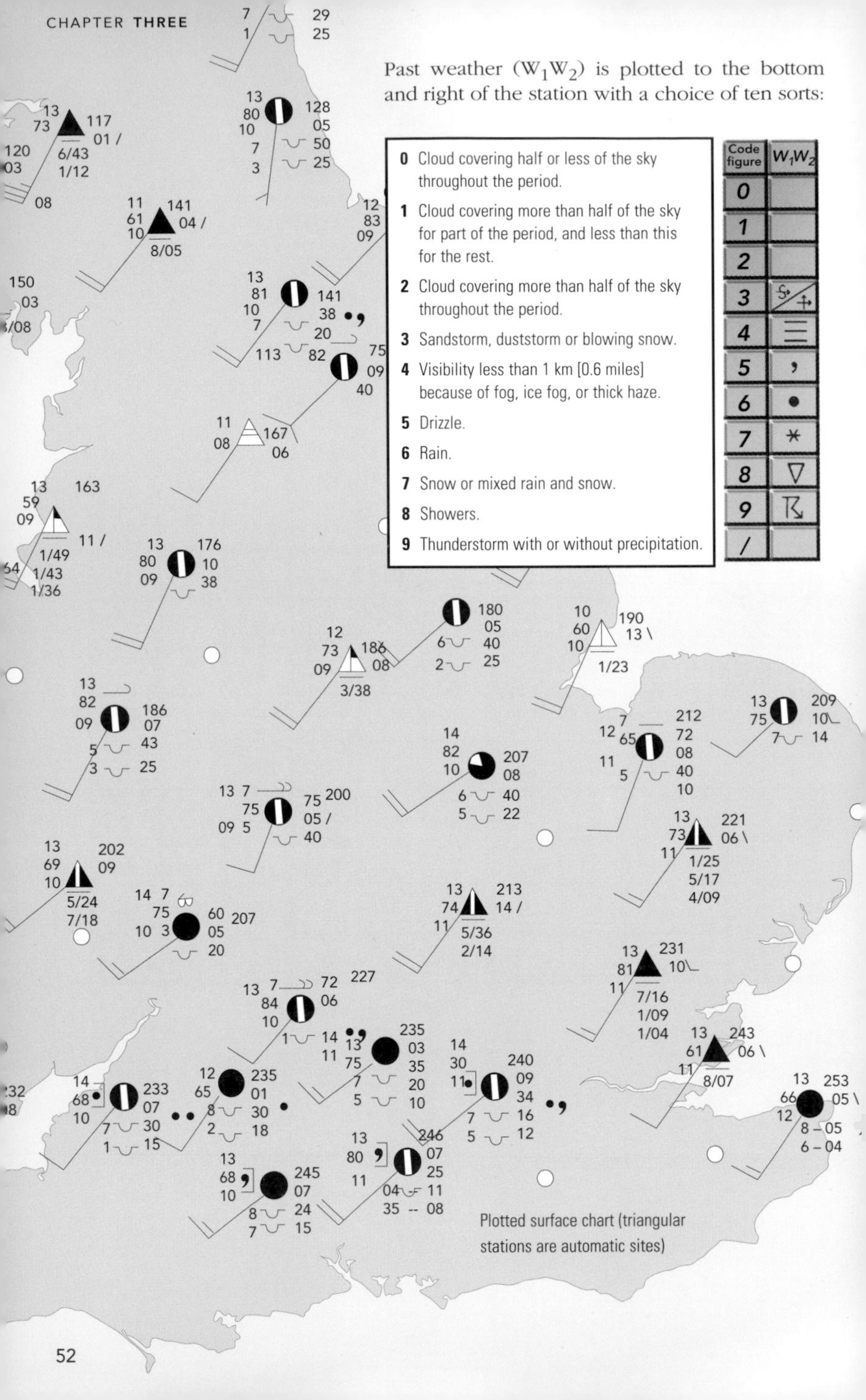

Plotted surface chart (triangular stations are automatic sites)

Analyzing and interpreting the weather map

Isobars

Modern weather services provide automatically plotted surface charts, upon which the analyst (or a computer-driven machine) will draw a variety of lines. The main features of interest in middle latitudes are the traveling lows, highs, and fronts that produce the varied weather across such regions.

Often the first lines to be drawn are the isobars, which join points of equal mean-sea-level pressure, because this determines the location and intensity (by the central value and surrounding pressure gradients) of the weather-producing disturbances: the lows, the highs, the troughs, and the ridges. By comparing successive charts, the forecaster can determine how they are moving and evolving.

The terms "low", "high", "trough", "ridge", and even "col" are evocative of the features shown on topographical maps of the land surface. Indeed, this is why the terms are used; the isobaric map can be thought of loosely as a topographical map, with troughs as "valleys" and ridges as "ridges"! An atmospheric col is the equivalent of its topographic namesake, too – a region of weak pressure gradient between two ridges and two troughs. Highs and lows may be many hundreds, or even a few thousand kilometers across.

Common features of mean-sea-level pressure charts

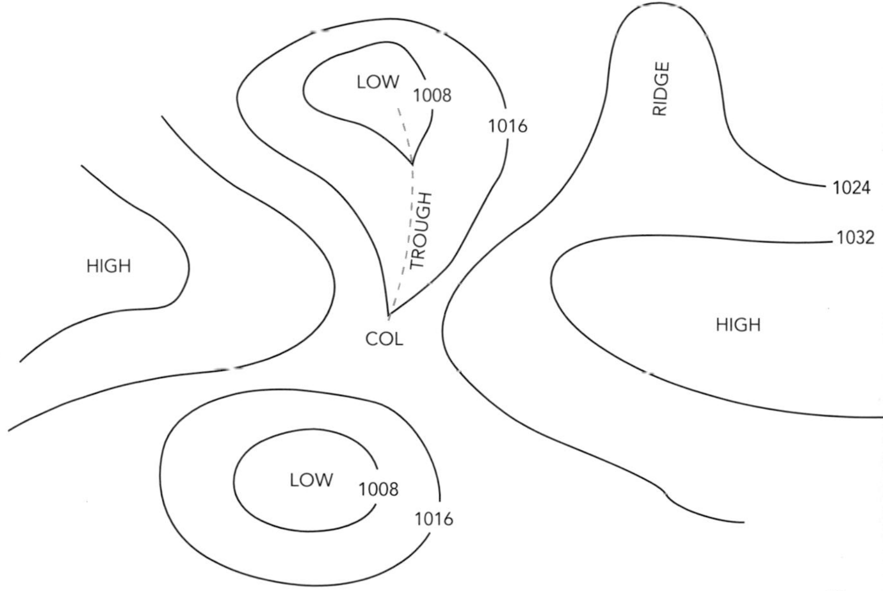

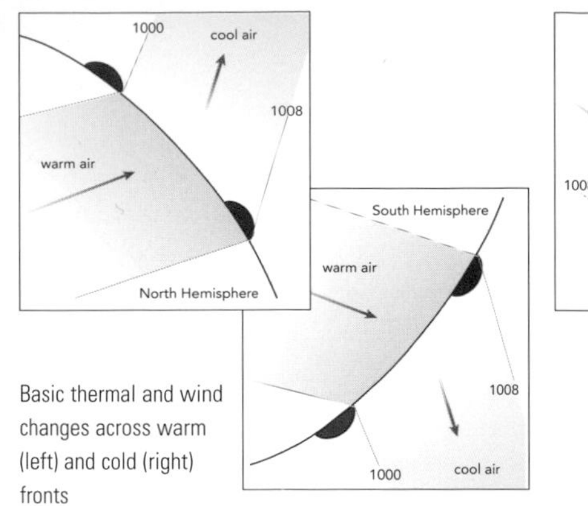

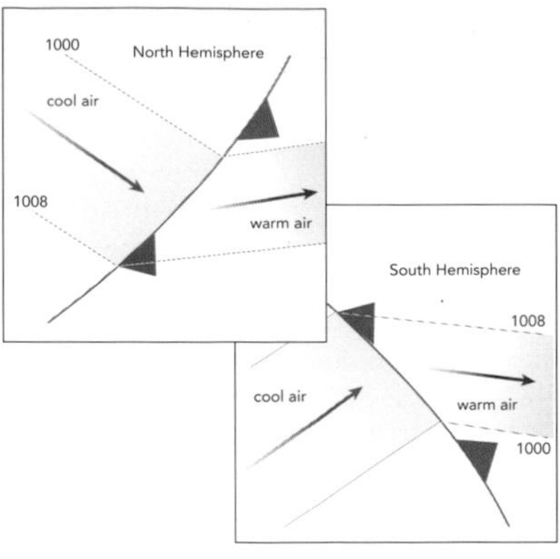

Basic thermal and wind changes across warm (left) and cold (right) fronts

Producing an isobaric analysis not only maps these features, but given the strength of the pressure gradient and the alignment of the isobars, also provides valuable information about the surface wind strength and direction. Tightly packed isobars represent a steep horizontal pressure gradient and, thus, very strong winds.

Fronts

Isobars are relatively easy to analyze, whereas much more skill is required to spot fronts on a surface chart. These very important weather features in middle and higher latitudes are shallow sloping zones that separate extensive air masses that have different values of temperature and humidity for example. Typically, a front slopes at about 1 in 100, cold fronts being somewhat steeper, and warm fronts somewhat shallower. Normally, a front is about 1 km [0.6 miles] deep, which means that it intersects the surface across a region some 100 km [60 miles] wide. Therefore, the weather changes associated with a front do not normally occur instantly, but gradually over a transition zone.

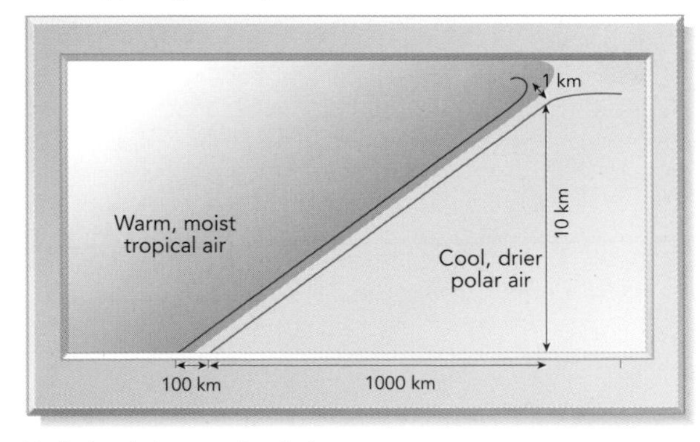

Idealized vertical cross-section of a front

Cold and warm fronts are the leading edges of cold and warm air masses that sweep generally toward lower or higher latitudes respectively. They are often indicated by a change in the orientation of isobars running across them, in association with a change of wind direction between the air masses that they separate. Broadly, the passage of a warm front brings warmer, moister air and a veering of the wind direction. This means that the wind shifts in a clockwise direction, typically over the space of an hour or so, from southeasterly to southwesterly in the northern hemisphere. In the southern hemisphere, the wind will shift from north-easterly to southwesterly, for example.

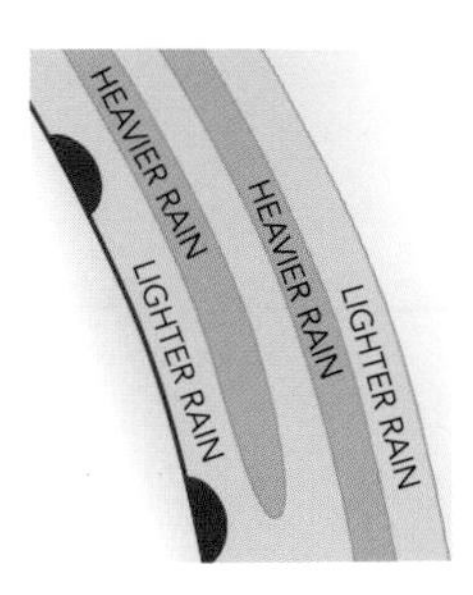

Typical rain region ahead of a surface warm front

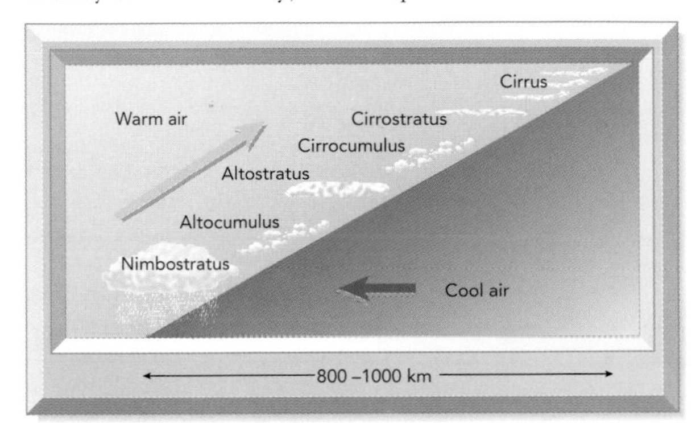

Idealized cloud sequence associated with a warm front

Ahead of the warm front

The area ahead of an approaching warm front is often influenced by the signs of the advancing warm, moist air as it glides across the sloping zone between the two air masses. In fact, much of the warm air streams beyond the line where the front meets the surface. High above the Earth's surface, the first signs of cirrus cloud will occur. This can be up to 600–700 km [370–430 miles] ahead of the warm front at the surface.

As the warm front approaches a point on the surface, the base of the cloud produced by the overrunning warm, moist tropical air gradually lowers. This is indicated by the gradual progression from cirrus to cirrostratus/cirrocumulus, followed by altostratus/altocumulus and a thickening into nimbostratus, that produces precipitation that reaches the surface.

Closer to the surface, the rain falling through the damp layer below the nimbostratus cloud will evaporate a little, cooling the air. Sometimes, this process leads to condensation in the damp air, creating "scud" or fractostratus clouds. These are the ragged low clouds that fly across the sky during conditions of strong winds and moderate or heavy rain.

Typical rain region ahead of a surface cold front

The leading edge of the warm frontal rain can occur some 200–300 km [120–180 miles] ahead of the surface front and cause a few hours of precipitation before the arrival of the warm sector air. Precipitation rates in this region would be about a few millimeters an hour, although radar observations reveal that rain bands often occur, with heavier, localized bursts.

The description given is of a typical warm front. In reality, however, each front is different. The cloud sequence may not follow the same pattern, and the warm frontal rain band may be less extensive, or slower moving, for example.

Broken stratus in a warm sector

The warm sector

The region between a warm and cold front is known as the warm sector. Quite often, it is characterized by extensive layer cloud that can produce persistently miserable conditions on exposed coasts, but may break up into pleasantly sunny conditions to the lee of hills.

The relative warmth, dampness, and cloudiness of a warm sector are an expression of the air's origin in oceanic regions of much lower latitudes. Precipitation within this sector is generally widespread; many frontal depressions exhibit a band of enhanced activity ahead of, and parallel to, the surface cold front. Visibility is often poor to moderate, and hill fog can be a problem in upland areas where the extensive low cloud has a base that is below the tops of hilly areas.

Behind the cold front

The passage of a cold front, as its name implies, usually produces a drop in temperature and dewpoint. This leads to cooler and drier conditions – in terms of absolute humidity or the amount of water vapor in the air.

The polar air that streams across the surface behind a cold front generally provides better visibility because it is often unstable, turning over in great depth and becoming well mixed. This instability also produces showery weather, precipitation being short-lived and falling from deep cumulus clouds. These characteristics are common over middle- and high-latitude oceans, but not over continental areas like North America. There, cold air that sweeps southwards from Canada in the winter does not experience significant surface heating over the cold land surface, and tends not to generate deep convective cloud.

In general, as a cold front passes, the wind veers from southwesterly to westerly or northwesterly in the northern hemisphere. In the southern hemisphere, it tends to shift from northwesterly to southwesterly.

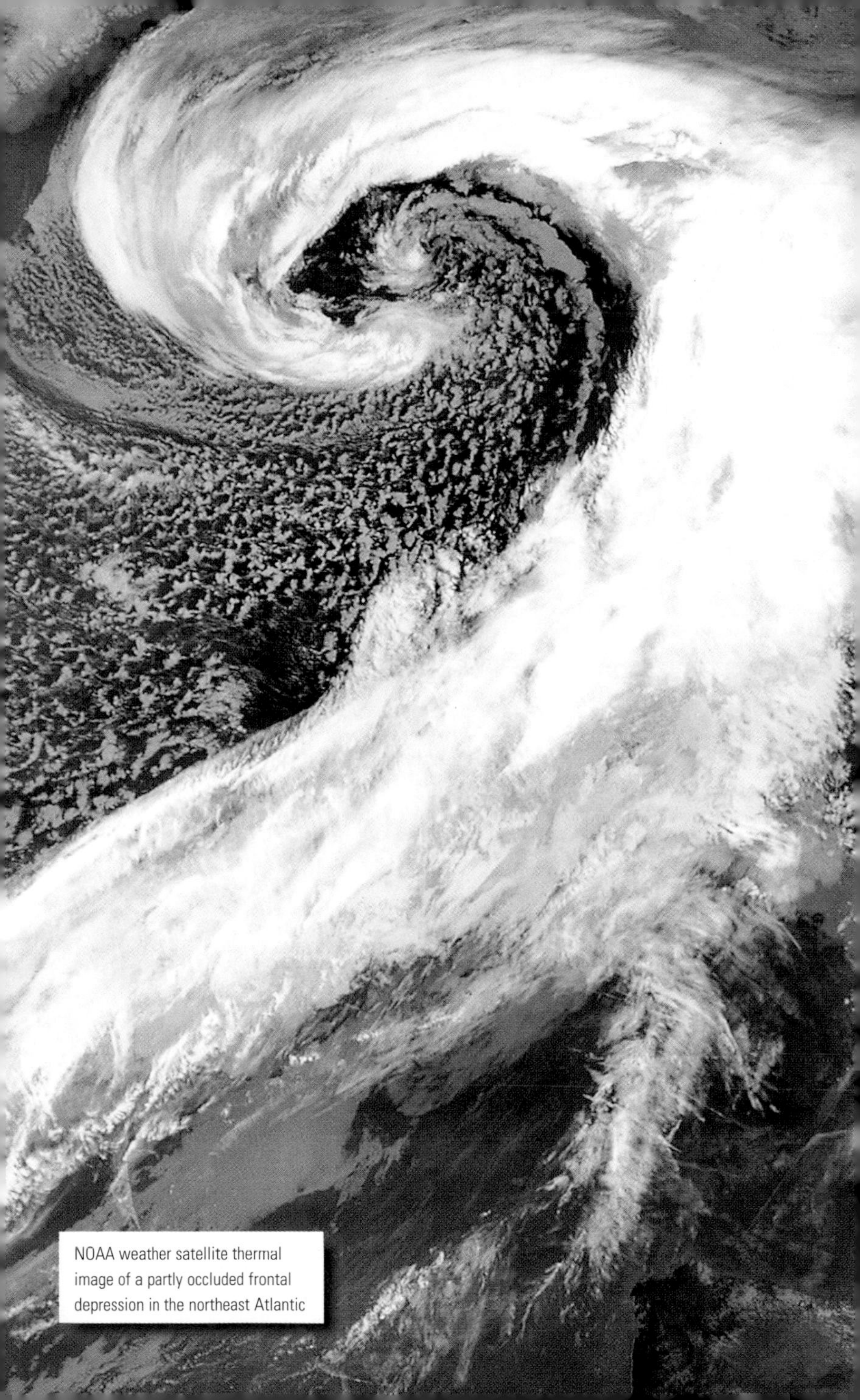

NOAA weather satellite thermal image of a partly occluded frontal depression in the northeast Atlantic

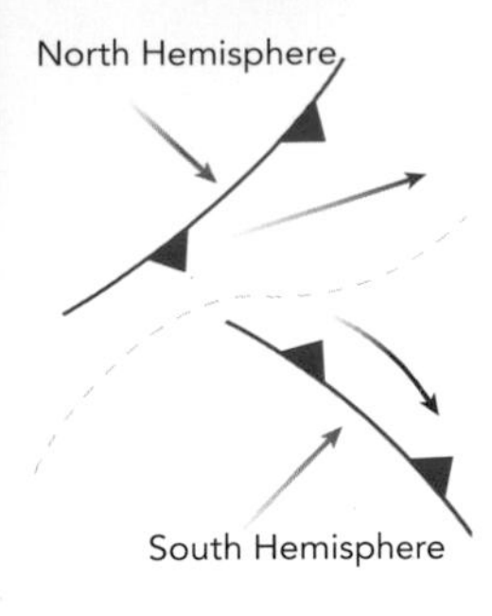

Typical wind change across a surface cold front

From satellites, it is possible to observe the patterns of cloud that form the "signatures" of these different regions of frontal depression.

Upper air

Weather observations gathered from radiosondes and other platforms, like satellites, are plotted on upper air charts to represent winds and temperatures, for example, at a variety of levels above the surface.

The data plotted on these charts are dry bulb and dewpoint temperature, wind direction and speed, and the height of the pressure surface above mean-sea-level (see the chart below). The observations can be analysed in various ways. Tightly packed solid contours on the chart, for example, indicate that very strong south-west winds were blowing above the central North Atlantic.

It is common for analysts to draw contours of the height of the pressure surface above mean-sea-level at, for example, intervals of 60 m [200 ft]. This produces a topographical map of the pressure surface which, like a surface weather map, displays ridges, troughs, and steep gradients. It is possible to calculate the wind speed from the steepness of the isobaric surface's slope.

Mapping the upper-air features illustrates the presence of the great snaking flow of the wind in the narrow jetstreams, and the slowly-evolving "long" or "Rossby" waves that are intimately linked to the intensity and motion of the surface weather.

Height of the 500 mbar surface (dm) above mean-sea-level (solid contours) and the vertical distance between 1000 and 500 mbar (dm) (dashed contours). Larger values of this "thickness" indicate a warmer layer, smaller values, a colder one

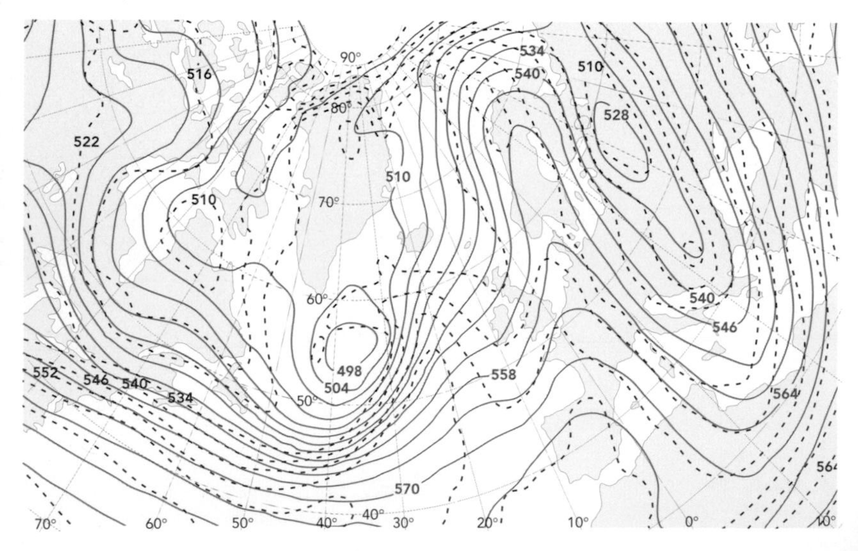

chapter four 4

Global Weather

It is important to understand the links between patterns of global pressure, circulation, temperature, humidity, and precipitation before investigating individual weather-producing disturbances, such as the frontal depressions of middle latitudes, or the violent hurricanes and typhoons of the tropics.

Averaging one weather station's time series of, for example, surface temperature, mean-sea-level pressure or wind speed produces a summary of the impact of transient weather features at that particular location. When these values are mapped across the Earth, they reveal the large-scale average patterns of weather phenomena that span weeks, months, seasons, or even years.

Mean-sea-level pressure

Northern winter and southern summer

Averaging the mean-sea-level atmospheric pressure for a number of Januarys and Julys summarizes the mean location and intensity of lows, troughs, highs, and ridges during the extreme seasons. There are major features during a typical northern winter/southern summer that express the common presence of disturbed and settled weather.

January mean-sea-level pressure (mbar)

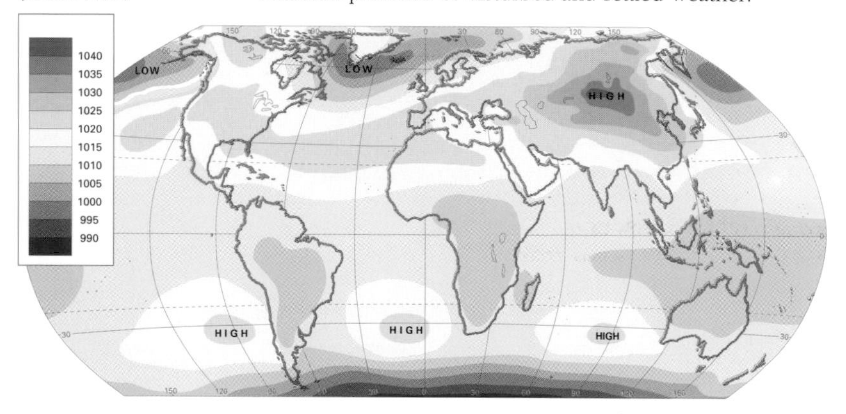

The middle-latitude cyclones

The Iceland Low and Aleutian Low, which occur in the higher latitudes of the North Atlantic and Pacific Oceans respectively, are reflections of the traveling low-pressure systems that run typically from southwest to northeast across these oceans during the winter. The minimum pressure values mark the point where, on average, the depressions, or cyclones, reach their deepest (lowest). The southwest/northeast alignment of their troughs indicates the mean track of the depressions in the winter. The map indicates that the long-term average value across the centers of the Iceland and Aleutian Lows is around 995–1000 mbar.

The trough that stretches northeastwards from the former is more extensive than that linked to the Aleutian Low. This is largely an expression of how the traveling cyclones

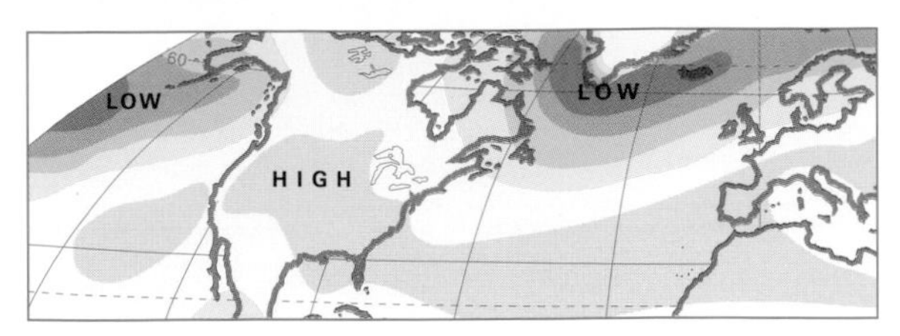

January mean-sea-level pressure (mbar)

are able to penetrate quite deeply into the Arctic Basin via the broad Norwegian Sea, in contrast to the more limited poleward excursions across the Bering Strait.

There is no particular value that makes an area of low pressure "low". It is simply a minimum value that occurs over the region of interest. The map indicates that the pressure features are quite large, typically a few thousand kilometers across. A low could be as little as, perhaps, 930 mbar in a very deep Atlantic depression on a winter's day, or as "high" as 1015 mbar, in a summertime "heat low" over land.

Continental anticyclones

In contrast to the maritime lows in the winter hemisphere, the extensive cold continents are marked mainly by the presence of very large highs, or anticyclones. The centers of these two major features lie deep in the middle-latitude continental interiors of Asia and North America. The most intense is the Asian High, with a long-term central value above 1040 mbar; the center over the United States is less intense (1020–1025 mbar), but, nevertheless, it has a strong influence on the regional weather.

The highs are the products of intense radiative cooling that occurs across these vast land masses in the winter. As with lows, there is no specific value of pressure that defines such a feature as a "high" – the pressure quoted is simply the maximum value that occurs across an extensive region.

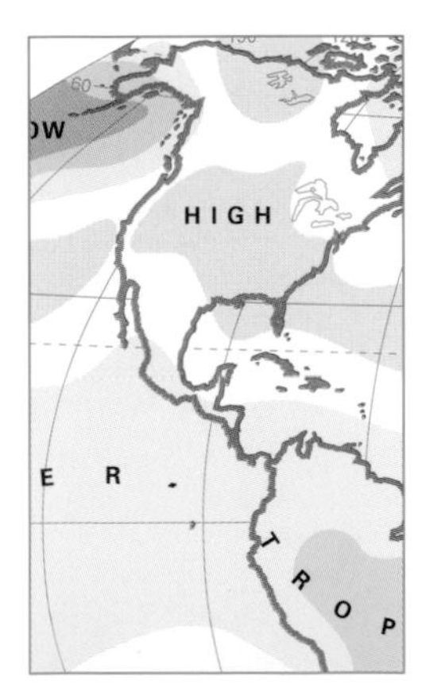

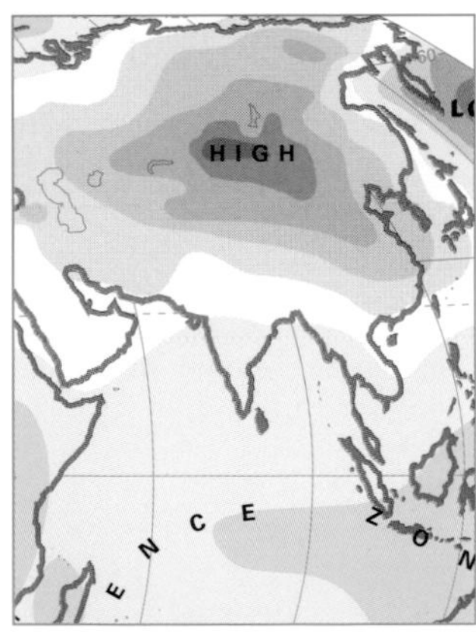

January mean-sea-level pressure (mbar)

Therefore, a high could have a value of, perhaps, 1055 mbar on a particular day, and 1015 mbar on another. The important aspect is that it is the largest value observed, with pressure gradually increasing toward it across a few thousand kilometers in a typical anticyclone.

Subtropical oceanic anticyclones

In addition to the cold continental winter anticyclones, regions of high pressure occur across the subtropical North Atlantic and Pacific Oceans. These are the Azores and Hawaiian Highs, which dominate the weather in these regions. They are much warmer than their continental counterparts and are much deeper, stretching throughout the depth of the troposphere. Cold anticyclones are quite shallow, being recognizable as highs only up to 1.5–2 km [approximately 1 mile] above the surface.

The Equatorial Trough

Moving toward the equator from the subtropical anticyclones reveals a broad low-latitude minimum of pressure known as the Equatorial Trough. This is something of a misnomer, since the feature occurs quite far south, across the intensely heated southern continents of South America, Southern Africa, and Australasia.

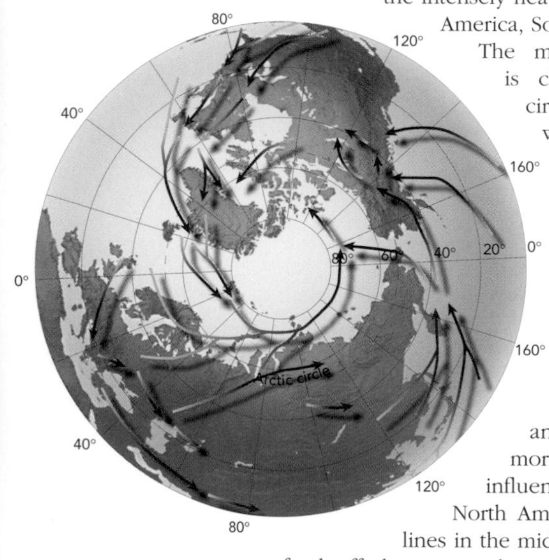

Major Northern Hemisphere depression tracks in January

The middle-latitude Southern Ocean is characterized by an elongated circumpolar belt of low pressure, which is virtually unbroken, unlike the northern lows. Its presence is a reflection of depressions that travel unimpeded around the open Southern Ocean, providing the strong westerly winds associated with the Roaring Forties that skirt the Antarctic continent.

In contrast, the North Atlantic and Pacific storm tracks run much more south-west/northeast, being influenced by the alignment of eastern North American and eastern Asian coastlines in the middle latitudes. These depressions feed off the strong thermal contrasts that exist between the northern hemisphere's continental and oceanic regions: the temperature gradient separating them has the same orientation.

There are warm subtropical anticyclones located over the South Pacific, South Indian, and South Atlantic Oceans, which give way northwards to the Equatorial Trough. There are no continental highs because these regions are strongly heated in the summer and are characterized by this shallow low-pressure (1005–1010 mbar) feature.

Polar regions are subjected to seasonal changes of pressure. The Arctic tends to experience a weak high in the winter, and a shallow low in the summer. The Antarctic is more of a problem in that continent is so high in general that reducing the pressure values to sea level becomes unrealistic. Broadly, it experiences relatively high pressure throughout the year.

A significant contrast between the northern and southern hemispheres that affects the pressure patterns is that in the former, the continents widen toward the pole (and surround the Arctic Ocean), while in the latter, they taper toward the pole, giving way to the circumpolar ocean that surrounds the massive continent of Antarctica.

July mean-sea-level pressure (mbar)

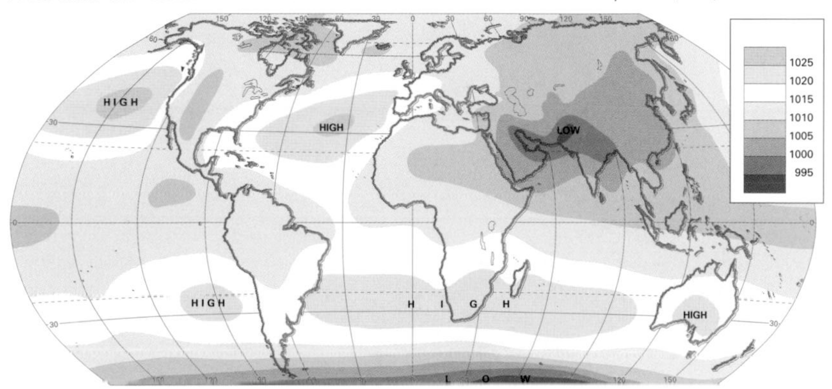

Northern summer and southern winter

The centers of low pressure, which are so marked over the northern oceans in January, are weaker or barely discernible in July, having shifted polewards. The extensive continental anticyclones are replaced by large-scale low-pressure features. Over Asia, this change is marked by a depression that lies across western India and Pakistan. The switch over this continent from extensive high to extensive low pressure is linked to the evolution of the monsoon from its winter to summer phase.

The summer hemisphere subtropical highs intensify or become higher. Both exhibit an increase of some 5 mbar and migrate a few degrees of latitude northwards. As in winter, the east–west continental/oceanic pattern in this hemisphere is caused by the break-up of the major pressure features into very large highs and lows. Moving towards the equator from these subtropical highs leads to the Earth-girdling Equatorial Trough that, over continents especially, exhibits a substantial seasonal migration.

The subtropical anticyclones in the wintertime southern hemisphere form a virtually complete belt, while the circumpolar lows still occur around the Antarctic with noticeably low pressure values.

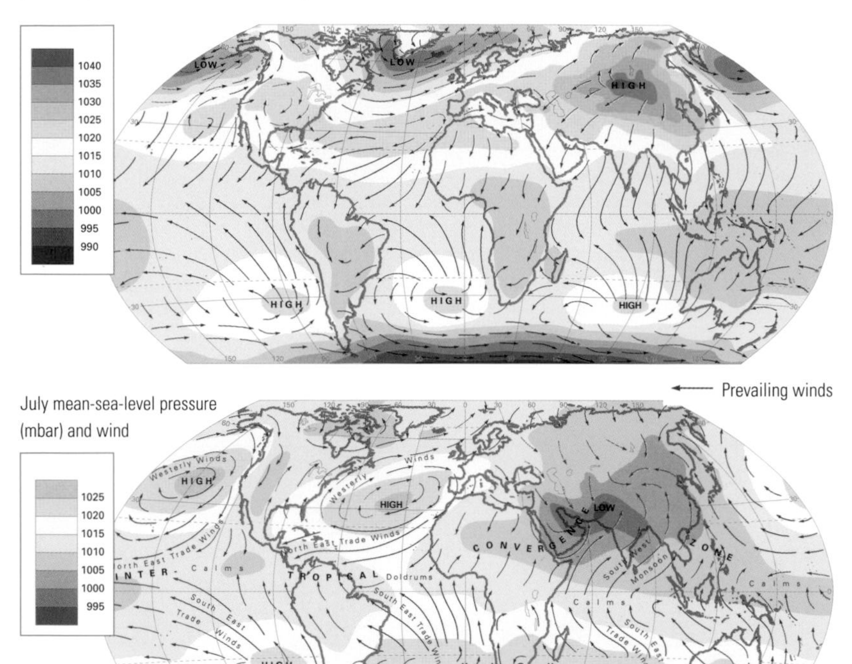

January mean-sea-level pressure (mbar) and wind

July mean-sea-level pressure (mbar) and wind

Wind

Northern winter and southern summer

The relative locations of the highs and lows in January and July determine the pattern of prevailing winds and, in part, the nature of the weather experienced around the Earth. In January, for example, the North and South Atlantic and their surroundings are influenced by the important source regions of air: the two subtropical anticyclones. It is clear that, at the surface, the winds flow clockwise out of these in the northern hemisphere, and counterclockwise in the southern hemisphere.

Parts of these outflows run toward the equator from both highs as the North-East and South-East Trades. Together, these culminate in the Intertropical Convergence Zone (ITCZ).

The Trades and the Intertropical Convergence Zone

The Trades are known for their strength and constancy over the tropical oceans, but they slow dramatically as they converge toward each other and enter the ITCZ. This feature is most noticeable over the oceans, and is typified by the infamous light and variable winds of the Doldrums. It is also well known for very strong ascent caused by the surface convergence of the hot, humid Trades; this shows up as cloud clusters that produce many thunderstorms in the ITCZ. Thus it is a feature of global significance, especially as it supplies water for much of the tropical world.

The Meteosat full-disk visible image illustrates a typical northern winter location for the ITCZ. It stretches around low latitudes and dips furthest south into the heated summertime southern continents, producing a significant seasonal maximum of rainfall. While the average wind flow for February points to a "smooth" ITCZ, the day-to-day pattern can be very different, with cloud and rain-bearing disturbances presenting a complex picture.

Meteosat visible image on a January day

On the eastern and western flanks of the subtropical anticyclones, the air flows generally parallel to the adjacent coasts, but also penetrates into the southern continents to supply the ITCZ.

Mean location of the ITCZ in February

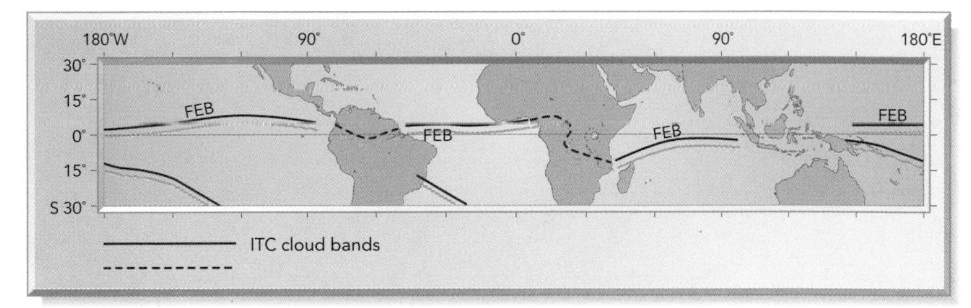

The middle-latitude westerlies

On the poleward sides of the anticyclones, major warm and moist currents of air move toward the poles as southwesterlies and north-westerlies in the North and South Atlantic respectively. Across the British Isles and western Europe, the southwesterly wind direction predominates. This maritime stream of air contrasts strongly with that on the other side of the North Atlantic, which affects Labrador, the Maritime Provinces and northeastern USA. Here, the prevailing wind direction is northwesterly, between the Iceland Low and the high over the USA. This means that most often the air comes from the cold, dry regions of North America's higher latitudes.

Thermal infrared image of an elongated frontal cloud band (center to top right) and cold air flowing off a springtime USA (top left): clear on land; showers over the western Atlantic

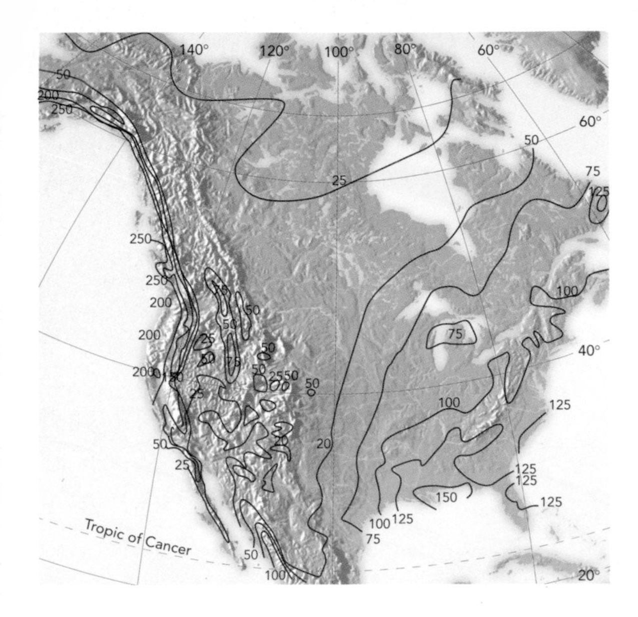

Mean annual precipitation (cm) over western North America

In a similar fashion, mild and moist south-westerlies flow toward southern Alaska and western Canada. The situation here differs from that in the northeastern Atlantic because of the Rocky Mountains. These are aligned more or less at right angles to the tracks of the traveling lows, whereas Atlantic depressions can move right across the lower land of northern Europe unimpeded. Much of the precipitation from the north-eastern Pacific depressions is deposited on the Rockies, to the detriment of the arid high plains to the east. In fact, the dryness of this region is related to the presence of the extensive rain-and-snow scavenging mountains to the west.

On the western flank of the Aleutian Low, the predominant flow is from the north-west, from very cold stretches of Russia and northern China.

During summer in the southern hemisphere, the westerlies blow parallel to the lines of latitude, between the subtropical highs of the South Atlantic, South Indian, and South Pacific Oceans and the higher-latitude low-pressure belt.

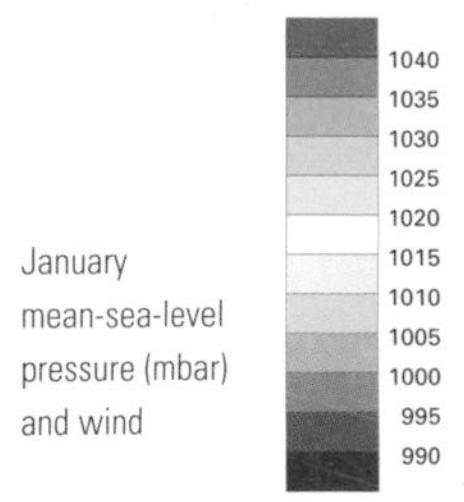

January mean-sea-level pressure (mbar) and wind

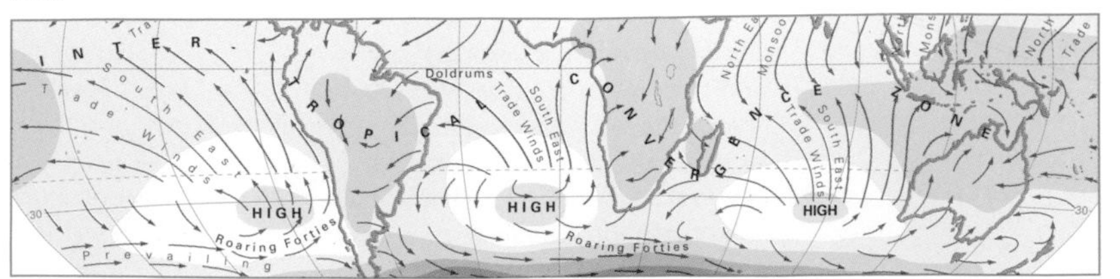

Continental anticyclones

The Asian region's winds are dominated by flow from the wintertime high into the Aleutian Low systems. Either they blow into the ITCZ as the North-east Monsoon, or toward the Arctic Ocean. Although the North American High is important, it does not dominate such an extensive region.

Northern summer and southern winter

The Trades and Intertropical Convergence Zone

The Trades exist throughout the year, with very marked and important migrations north and south of the ITCZ, particularly over the tropical continents. The ITCZ reaches its northernmost limit during the height of the northern summer, its most "famous" excursion being across southern Asia. The convergence zone and associated low center and troughs contrast markedly with the January pattern of extensive north-easterlies across this region. This seasonal change in the surface winds across India, for example, is the signature of the Monsoon, which takes its name from an Arabic word meaning "reversal".

Similar, but less extensive, seasonal wind reversals also affect the southern part of West Africa, the southwestern USA and northern Mexico.

The middle-latitude westerlies

These still occur in the northern ocean basins, but generally are less extensive and less vigorous than in the winter. The subtropical highs on the westerlies' southern flank intensify into the summer and shift slightly polewards. They still supply the Trades which, as in January, are most significant across the tropical oceans. The Roaring Forties of the Southern Ocean persist virtually all year, blowing powerfully between the oceanic subtropical highs and the circumpolar low-pressure region.

Meteosat visible image on an August day

Continental anticyclones

Apart from Antarctica and Australia, the less extensive continents of the southern hemisphere do not have substantial highs associated with them. Only Australia is large enough in the subtropical/middle latitudes to produce an anticyclone and, thus, to influence the mean wind pattern regionally.

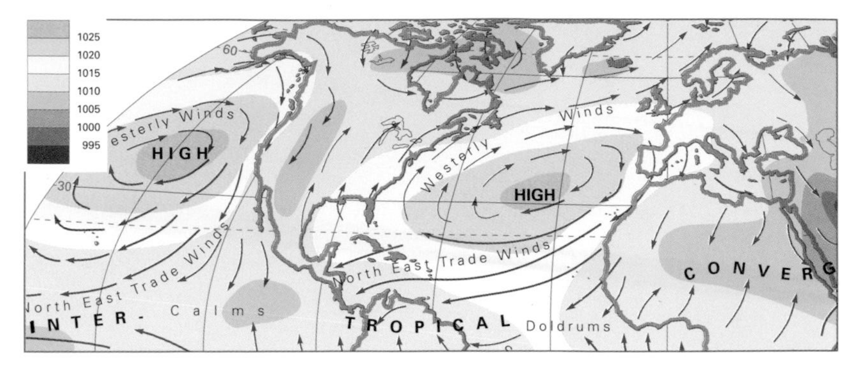

July mean-sea-level pressure (mbar)

Vertical wind patterns

So far, the winds under consideration have been those blowing across the Earth's surface. The flow from surface highs into surface lows implies that there must be some connection through large-scale vertical movements of air.

By considering the average wind flow patterns at various horizontal levels within the troposphere, meteorologists can deduce the associated pattern of vertical motion that connects a high with a low. The result illustrates the strong ascent linked to the very extensive swirling of air as it enters a surface low, and the deep descent that supplies the air blowing out of a surface high.

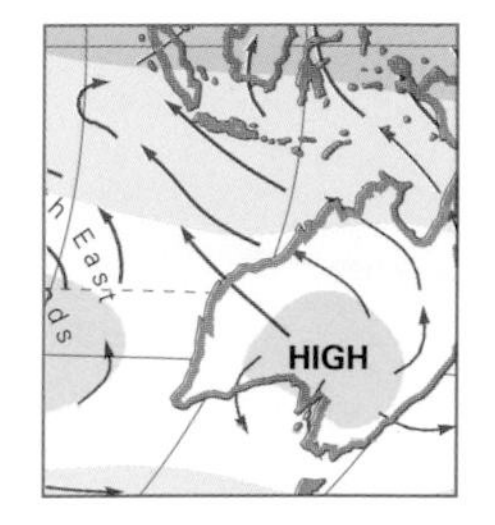

July mean-sea-level pressure (mbar)

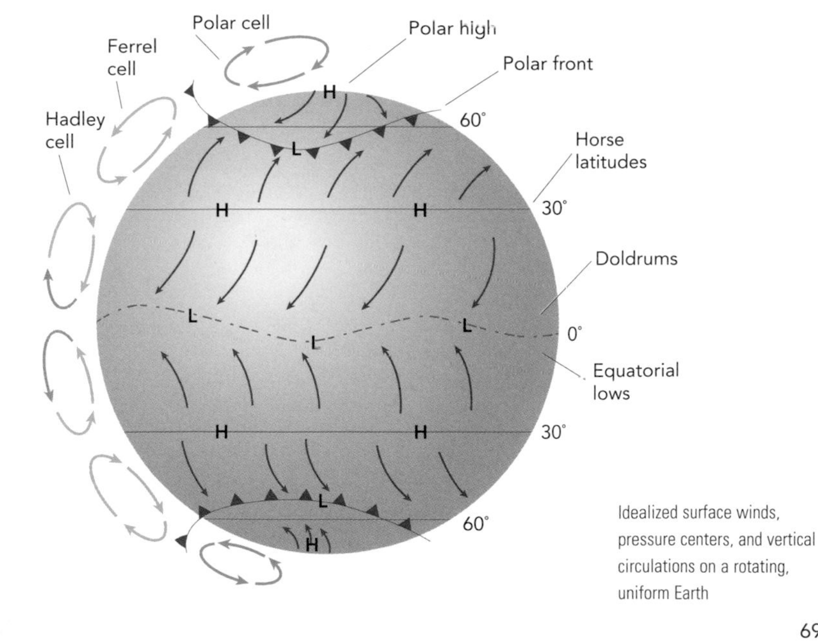

Idealized surface winds, pressure centers, and vertical circulations on a rotating, uniform Earth

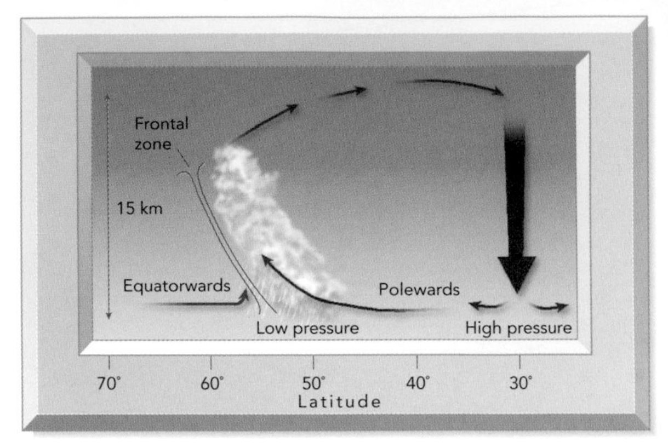

Basic model of a Ferrel cell

Hadley cell

Each hemisphere displays three distinct types of vertical air circulation, known as cells. The deepest, most powerful and most extensive is the Hadley or tropical cell, named after the English scientist George Hadley, who suggested its existence in the first half of the 18th century. It comprises:

1. Generally very vigorous, tropical thunderstorms (known as "hot towers") associated with the low-level convergence of warm, moisture-laden air in the ITCZ. This is a very significant region of heavy precipitation that supplies many tropical locations with life-giving seasonal rains.

2. Flow from the top of this thundery zone, in the upper troposphere, toward both poles. This air gradually cools as it moves polewards and, at around 30°N and 30°S, sinks through the troposphere. This region of deep subsidence is linked to the subtropical highs and generally very dry weather.

It is no coincidence that most of the world's hot deserts occur beneath this region of persistently subsiding air.

3. Surface return flow from subtropical highs toward the ITCZ as the North-east and South-east Trade Winds. These strong winds pick up vast amounts of water vapor by evaporation from the warm tropical oceans over which they flow. This increasingly humid air converges from each flank into the ITCZ, where there is vigorous ascent within the massive thunderstorms that characterize the zone. Therefore, most

of the rain that falls from these cumulonimbus clouds comes from evaporation over the tropical oceans.

Ferrel cell

Much weaker than the Hadley cell, the Ferrel or middle-latitude cell is named after the 19th-century American meteorologist William Ferrel. It comprises:

1. Low-level currents of air that flow polewards from the subtropical highs. This warm, moist air forms the warm sectors of middle-latitude frontal depressions.

2. A region of rising air, at around 50–60°N and 50–60°S, represented by the large-scale rise of the warm, moist air in frontal depressions. This is the flow of tropical maritime air across gently sloping warm fronts. These common frontal depressions are the major source of precipitation for middle- and higher-latitude areas of the world.

Like the ITCZ, this is a region of convergence at the surface, where air flows together from different directions, ascending to produce cloud and often widespread rain, drizzle and sometimes snow. Unlike the ITCZ, however, the winds carry less moisture. Those from higher latitudes are particularly cool and quite dry.

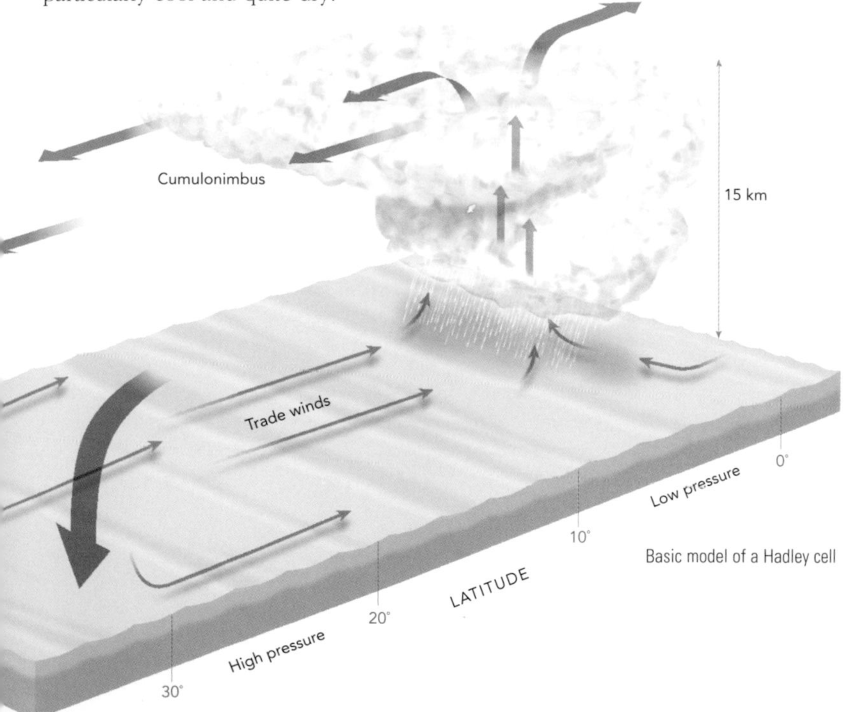

Basic model of a Hadley cell

3. A return flow in the upper troposphere that heads toward the upper outflow from the ITCZ. These two flows converge above the subtropical highs, and are matched by the two currents that diverge directly below them at the surface.

Polar cell

The final component of vertical air circulation is the weak Polar cell. This comprises:

1. Gentle sinking in the highest latitudes, associated with surface highs.

2. Surface flow toward the equator, some of which ultimately undercuts the warm-sector tropical maritime air in the frontal zone. This forms the leading edge of the polar air behind a cold front.

3. A weak return flow from above frontal depressions towards the poles.

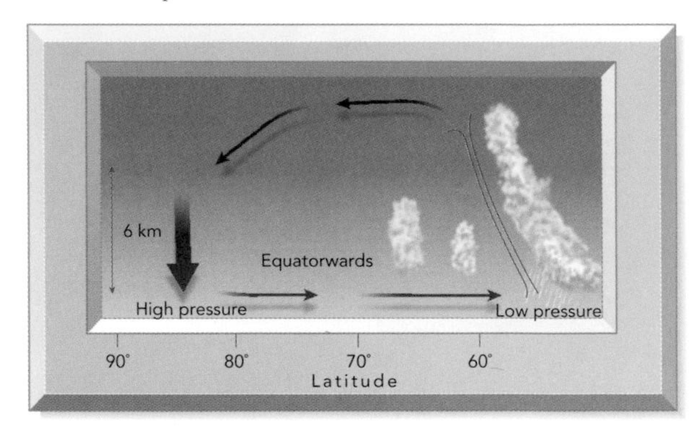

Basic model of a Polar cell

Surface temperatures

There is a great seasonal variation in the amount of solar radiation received at the surface, especially at the higher latitudes. The tropics do not see such a large change. These important differences are mirrored in the seasonal change of ocean surface temperature and the temperature over the continents.

January

Consider first the difference in air temperature between a couple of equatorial locations and two high-latitude sites, using the mean air temperature for January.

The January average temperature over Congo, at 15°E on the equator, is around 25°C [77°F], while at the same longitude in northern Norway it is about –10°C [14°F]. Now compare a site in Amazonian Brazil, at 60°W on the equator, with one in northwestern Greenland at the same longitude. The average temperatures at these places are 25°C and –30°C [77 and –86°F] respectively.

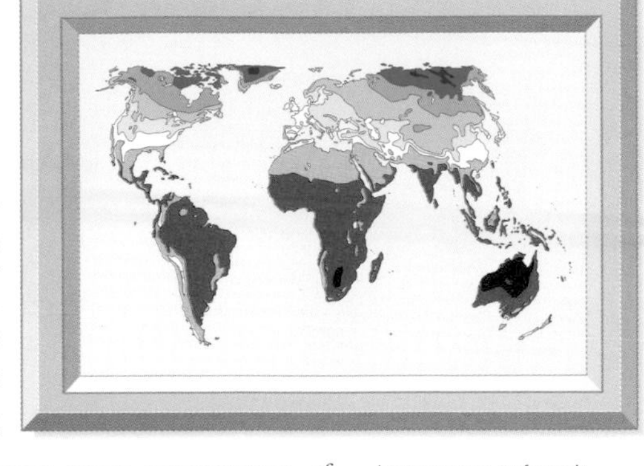

January mean surface air temperature (°C) over land

30°C
20°C
10°C
0°C
–10°C
–20°C
–30°C
–40°C

Cold pole

Therefore, in January, we would expect to see a temperature difference between the equator and high latitudes of about 35–55°C [95–130°F] in the winter hemisphere. The coldest temperatures in January are not centered on the pole, but typically occur in northeastern Russia. The Siberian city of Verkhoyansk experiences a January mean temperature of –40°C [–40°F], and on 6 February 1933 recorded the lowest northern hemisphere minimum of –68°C [–90°F]!

Mild west, cold east coasts

It is clear from the temperature patterns that in middle and high latitudes, the broad continents of the northern hemisphere are intensely cold in the winter. There is an east–west difference, however, with their eastern flanks being colder than the western. This is true of both Eurasia and North America. The coldest conditions on the east coasts are caused by the prevailing winds that blow off the cold continents. In contrast, the west coasts experience milder conditions partly because of the tropical maritime air borne by the traveling frontal lows that approach from the ocean. These are complemented by warm ocean currents that stream toward the western coasts. Both the Gulf Stream/North Atlantic Drift and the Kuro Siwo/North Pacific Current are crucial in this respect.

Baking Australia

During January, on average, the warmest places lie in the interior of South Africa and more widely over much of northern Australia, where the mean temperature is higher than 30°C [86°F]. Darwin tends to be the hottest state capital in Australia, having a mean maximum in January of 31.7°C [89°F]. The middle-latitude and maritime capital Hobart, in Tasmania, can only muster

January mean air temperature (°C) over land

21.5°C [70°F] as its average. Interestingly, the record summer maximum in any Australian state capital belongs to Adelaide where, on 12 January 1939, the high reached 47.6°C [118°F] ! It is the outbreak of extremely hot air from the interior that leads to these baking temperatures.

Ocean flow

The principal warm ocean currents of the southern hemisphere are known as the Brazil and Agulhas Currents. The extensive region of warm water that washes the shores of eastern Australia is also significant. The western flanks of the southern continents are influenced by cold oceanic flows toward the equator in the form of the Peru (or Humboldt), the Benguela and the West Australian Currents.

In comparison to the northern hemisphere's ocean circulation, the southern oceans are markedly cooler, being partly influenced by the cold Antarctic Circumpolar Current at high latitudes. The main cold currents in the northern oceans are the California, the Labrador, the Canaries, and the Oya Siwo.

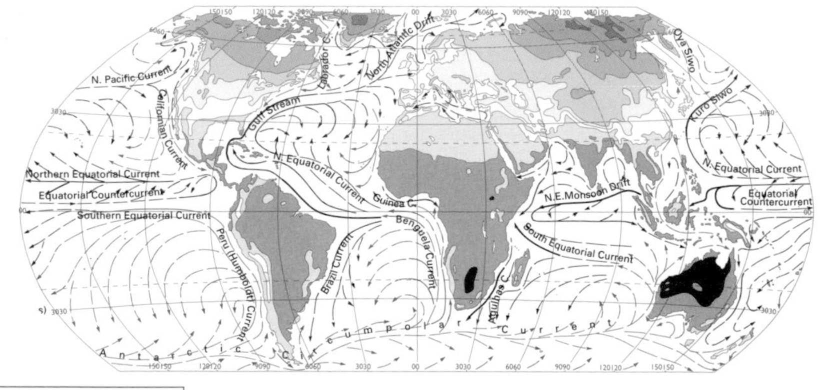

OCEAN CURRENTS

Cold	Warm	Speed (knots)
←--	←--	Less than 0.5
←	←	0.5 – 1.0
←	←	Over 1.0

Major surface currents in January

July

In July, the temperature contrasts between the equator and higher latitudes are quite different from those in January. The mean temperatures at the same points in Congo and Amazonian Brazil are still around 25°C [77°F], while the matching values for northern Norway and Greenland are about 10°C [50°F] and 0°C [32°F] respectively. This means that the differences are between 15°C [60°F] and 25°C [77°F]– or something like a half of those seen in winter.

This change in the temperature gradient from low to high latitudes holds for both hemispheres and throughout the depth of the troposphere. The approximate doubling of the difference from summer to winter is mainly due to the dramatic change from polar day to polar night. During the long polar night, vast areas of the highest latitudes cool substantially, while during the long days of the summer, temperatures reach significantly higher levels.

Conservative tropics

In general, the tropics experience a very small annual temperature range between the warmest and coolest months, because there is not much variation in the amount of solar radiation received by them throughout the year. This is why we define different seasons within the tropics in terms of when it rains, rather than temperature changes. In contrast, in the extratropics, the winter is significantly colder than the summer, so we use the annual round of warming and cooling as a basic means of defining the seasons.

Record heat

In July, the hottest conditions are to be found on the continents that have extensive tracts with mean air temperatures above 30°C [86°F]. It is within the most extensive of these, across Saharan Africa, that the world's highest screen maximum, and highest mean annual screen temperature, were observed. Respectively, these are 58°C [136°F] on 13 September 1922 at Al Aziziya, just south of Tripoli in Libya, and 34.4°C [94°F] from 1960 to 1966 at Dallol, in the low-lying Danakil Depression that straddles Ethiopia and Eritrea.

As in the southern continents in January, in part these hot areas are related to the presence of the sinking air of the Hadley cells, and also to their distance from the influence of air flowing from the oceans. The deep sinking motion in the subtropics is characterized by extensive clear skies promoting very high temperatures during the day, and chilly conditions at night in the "hot" deserts of the world. In hot coastal regions, however, such as the Persian Gulf, overnight temperatures remain high because of the generally very humid air. The water vapor in this air absorbs some of the upward Earth radiation and re-radiates part of it back to the surface, acting as a form of insulator.

This is an example of the Greenhouse Effect, in which gases such as water vapor and carbon dioxide keep surface temperatures higher than they would otherwise be.

Record cold

At the other end of the scale, the world's lowest mean annual screen temperature of –57.8°C [–72°F] belongs to the Russian Antarctic station of Polus Nedostupnosti. A contributing factor to this amazingly low value is the extreme dryness that accompanies frigidly cold air. There is virtually no water vapor in the atmosphere above the wastes of the Antarctic and, therefore, no "greenhouse" contribution from that source.

Precipitation

The global pattern of annual precipitation is strongly related to those of pressure and wind.

In middle latitudes, widespread (spanning hundreds of kilometers) precipitation is generated by the ascent of warm, moist air over fronts that sweep across oceans and adjacent continents. Frontal rain and snow move with the frontal depressions that create them. In many regions of the middle latitudes much of the rain and snow is provided by these traveling disturbances. The areas of their activity vary seasonally: they tend to shift toward the pole in the summer.

These systems are responsible for a good deal of the precipitation along the extreme western flank of North America, from the Gulf of Mexico to the Maritime Provinces of Canada, and also across much of Europe, including the Mediterranean in the fall and winter. They also affect regions from China and Japan to Kamchatka in Russia, as well as over and to the west of the southern Andes. Also across southeastern South America in the autumn and winter to the southern flanks of southern Africa. Australia and New Zealand are affected during their cooler season. These depressions produce widespread precipitation over the middle-latitude oceans as well.

During the summer, over middle-latitude continents, strong surface heating leads to significant showery rain. This tends to be shorter-lived and heavier than the frontal type.

Wet and snowy mountains

Annual totals indicate that around 500–2000 mm [20–80 inches] of precipitation falls across Europe in a typical year. The largest amounts occur in mountainous regions, especially those that abut the Atlantic Ocean and provide the first landfall for the traveling frontal systems. Broadly similar totals occur across western North America, heavier falls being concentrated along the relatively narrow mountainous coastal zone from central California northwards. The steep rain/snowfall gradient across this strip reflects the impact these massive coastal ranges have on frontal precipitation and the rain-shadow region to the east, where totals are below 250mm [10 inches]. These mountainous regions face the onslaught of winter depressions and suffer the snowiest conditions. The world record annual snow total comes from Paradise Ranger Station on Mount Rainier in Washington State, USA, where 31.102 m [102 ft] fell between 19 February 1991 and 18 February 1992!

Dry areas

The pattern of dry areas (with less than 500 mm [20 inches] a year) can be related to a number of causes. At the highest

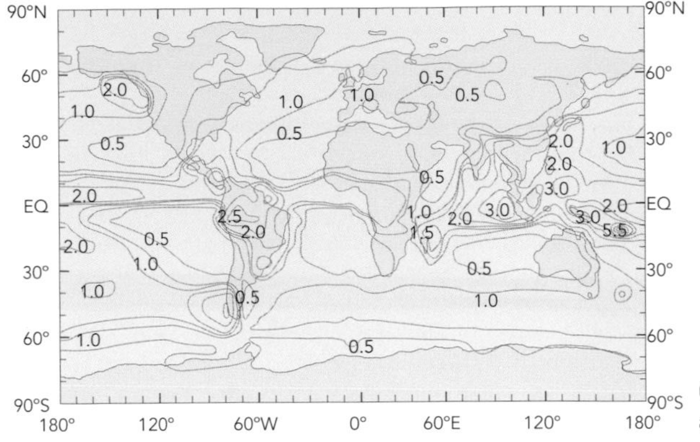

Observed and estimated annual mean precipitation (m)

of northern latitudes, low precipitation values are due, in part, to the low temperatures that prevail in those regions. The amount of water vapor contained in cold air is very small indeed. Even if the atmosphere provides a means of lifting the air to cool it enough for the water vapor to condense into clouds, little rain or snow is produced.

The aridity of the middle latitude continental interiors is partly due to a rain-shadow effect: for example, the high plains of the United States, and the dry region of western Argentina, in the lee of the Andes. Other areas, such as Siberia in Russia, which is east of the Urals and north of the Himalayas, are generally arid because, in winter, the massive

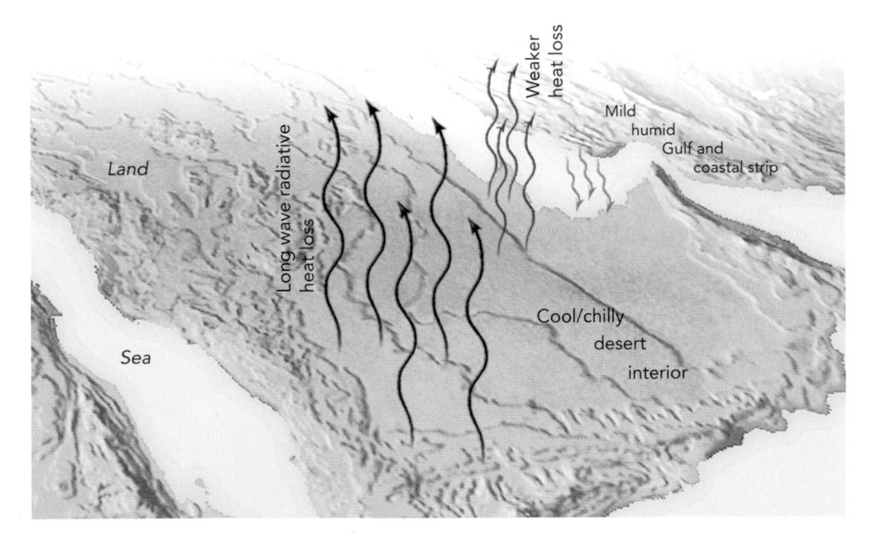

The impact of water vapor on night time cooling under clear skies

anticyclone suppresses any ascent and is very cold. In summer, surface heating will spark off scattered showery precipitation, but the region's remoteness from the sea means that very humid air rarely reaches it.

In contrast, the interior of the United States is exposed to very large incursions of moist air from the Gulf of Mexico. During the summer, these provide the essential ingredient of the torrential thundery downpours that can be linked to severe phenomena like large hailstones or even tornadoes.

The marked aridity of the Sahara, Arabia, and the Thar Desert of northwest India is essentially an expression of the sinking portion of the Hadley cell. The same is true of the Australian and Kalahari Deserts, and indeed the regions of scant rainfall that stretch across the eastern tropical/subtropical oceans.

Another group of arid land areas runs down western South America and Southern Africa. These are affected by the presence of cold ocean currents that flow toward the equator along the coasts. The Atacama and Namib Deserts are here, where the cold water suppresses any rain-producing ascent, but lies under extensive layer cloud just offshore. In fact, at Calama in northern Chile, no rainfall at all was reported in a 400-year period up to 1971!

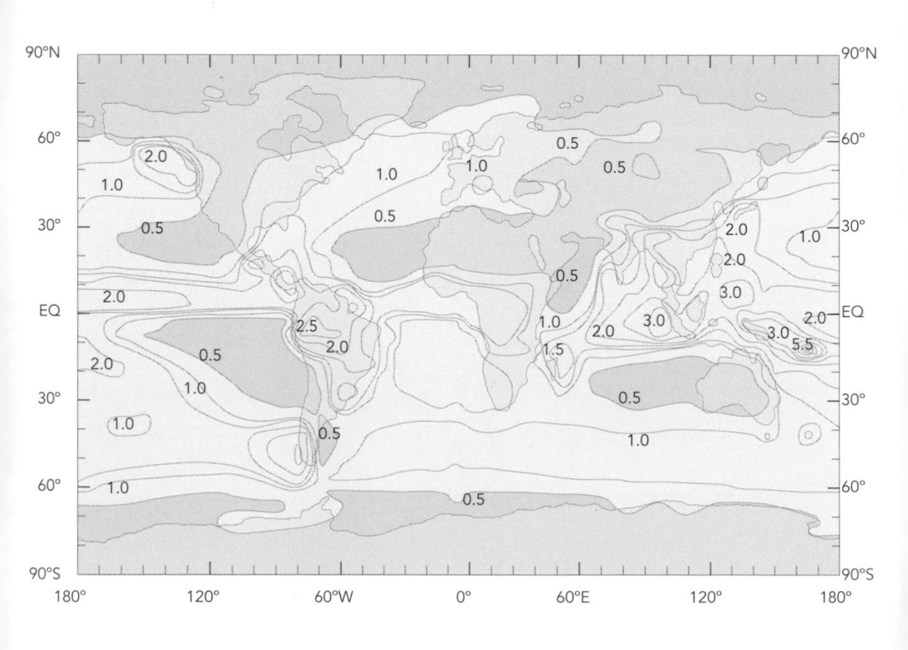

Regions where mean annual precipitation is less than 0.5m [1.6 ft]

ITCZ

The most extensive regions of precipitation are within the tropics. These areas are where the ITCZ and its deep convective clouds hold sway. Sometimes, hurricanes and typhoons are created within the regime of the Trade Winds, producing widespread heavy rain as well as their notoriously dangerous winds.

So, those places where 2000–3000 mm [80–120 inches] and more are observed are regions where the ITCZ is active, where mountainous coasts face onshore flow – particularly in monsoon areas such as western India and Sierra Leone, where hurricanes, typhoons or cyclones (Indian Ocean) run across land areas, and where mountainous islands like Indonesia trigger locally heavy showers.

Much of the heavy rainfall that falls across Southeast Asia and West Africa is monsoonal, while some of the large amounts over the Caribbean, Central America, the Phillipines, Vietnam northwards to Japan, and Madagascar are produced by intense tropical cyclones. These traveling rotating storms tend to be embedded in the larger-scale north and southeast Trades. Often, they are born on the eastern flanks of oceans and make landfall on their western flanks.

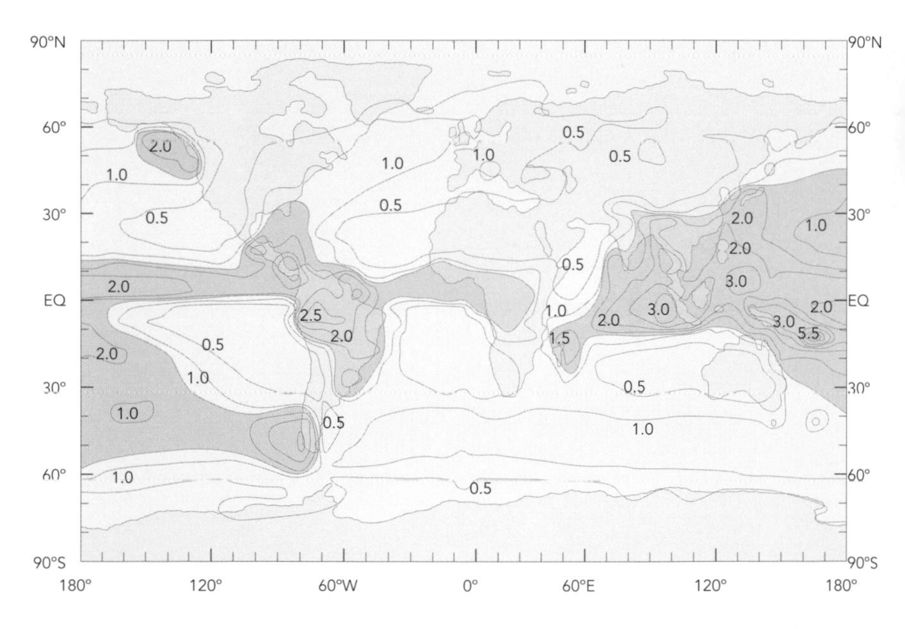

Regions where mean annual precipitation exceeds 1.5m [5 ft]

This annual pattern masks the seasonal migration of the ITCZ, so the heavy rainfall over South-east Asia occurs during the summer monsoon. Conditions during the winter monsoon are mainly dry.

Rainfall record

With the very high humidity levels within the oceanic tropics, and the intense surface heating there too, it is no surprise that the world's rainfall records are held by the region. The largest rainfall total for any 12-month period was recorded in Cherrapunji in the Indian tea-growing region of Assam. It returned an accumulation of 26.27 m [86.2 ft] from August 1860 to July 1861! In fact, the same station also holds the wettest month record: 2.93 m [9.6 ft] in July 1861.

The largest 24-hour fall comes from the island of Réunion in the western Indian Ocean, where an unbelievable 1.87 m [6.1 ft] was recorded during 15–16 March 1952!

chapter five 5

Explaining the Weather

Every day, we all experience one facet or another of the weather, even if this is only the view from a window on a dull day. Our modern insulated homes (mostly in urban areas), vehicles, and public transport systems make us less susceptible to the vagaries of the weather and, unfortunately, less aware of it, too. What makes the wind blow? How do clouds form and stay in the sky? How do they produce rain, snow, and drizzle? Why are there different types of fog? How do dew and frost form? These topics are discussed in this chapter.

Only on relatively rare occasions do we notice the impact of the weather. This may occur during a very windy and wet winter storm, for example, during a spell of prolonged dryness, on early-morning icy roads, or on a day blanketed by thick fog.

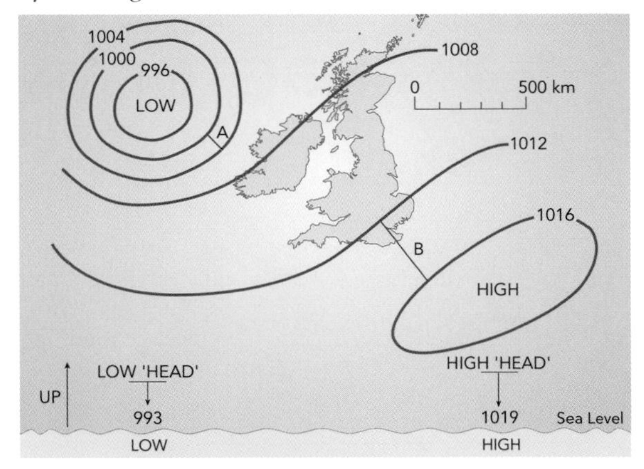

A: Steep pressure gradient (strong wind)
B: Gentle pressure gradient (weak wind)

Horizontal pressure gradient and wind

Pressure and wind

The difference between high and low pressure across the Earth's surface is the basic driving force that makes the air move. Using the extreme-season maps, for example, we can assess the difference in pressure between the centers of the Iceland Low and Azores High during the North Atlantic winter. This turns out to be about 25 mbar. By determining the distance between the two, which is something like 2500 km [1600 miles], we have the information necessary to calculate the horizontal gradient of pressure – or how fast pressure changes across the sea surface. In this case, it would be around 1 mbar per 100 km [60 miles].

A large pressure difference between a high and a low that are quite close will produce a steep gradient and a strong force to drive the air. A much weaker gradient, however, will generate less force. The steeper the gradient, the stronger the wind. The difference between the "head" of air above a high and that above a low drives the air from the high toward the low. This flow will decrease the mass of air in the column above the high, causing pressure to fall, and increase the mass of air in the region of the low, causing pressure to rise.

Spiraling winds

Although air flows from high-pressure regions toward low-pressure areas, the direction of the flow, or wind, is not straight from one to the other. The Earth's rotation causes the wind to be deflected so that it spirals out of the highs and into the lows. The direction of the mean prevailing winds across the Earth's surface in January and July are strongly related to the direction of the isobars. However, note that the winds do not flow exactly parallel to the isobars, but cut across them at an angle, illustrating their spiraling nature.

At the surface and through the lower troposphere, air spiraling out of an anticyclone must be replaced by air that is slowly sinking within the high. It follows that air spiraling into regions of low pressure must ascend. These rising and sinking motions are very important, for without them, there would be no weather as we know it.

Anticyclones are characterized by quite deeply sinking air, while cyclones exhibit large-scale ascending streams, often reaching great height. The extensive rising motion in the latter tends to produce cloud, wind and rain or snow, the ascent of moist air being nature's favorite way of producing these features. In high-pressure regions, however, the large-scale descent of air generally gives dry conditions, although quite extensive low layer cloud may also occur.

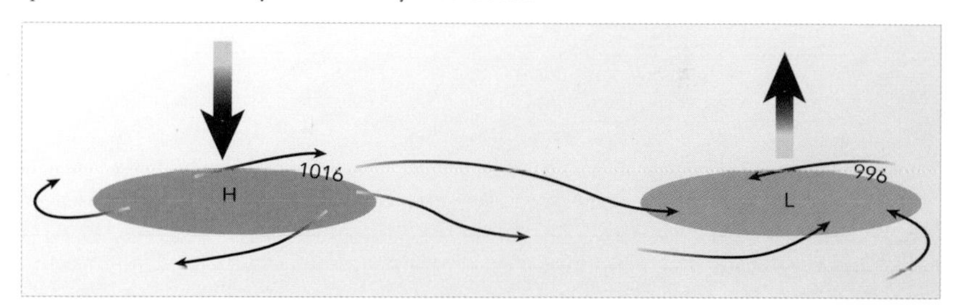

The link between descent/surface outflow and ascent/surface inflow

Clouds

All clouds are formed by the cooling of moist air down to its dewpoint temperature. At this stage, the air becomes saturated. Further cooling causes the water vapor to gradually condense out of the air as myriad cloud droplets.

The amount of water vapor contained in saturated air depends on the air temperature. Cold air is capable of holding small amounts, while very warm air can contain much more. This marked increase in the saturation value of water vapor with temperature means that moist, cold air generally produces much less precipitation than moist, warm air.

Near sea level, a cubic metre of air weighs about 1.2 kg [3 lbs], while one of water weighs 1000 kg [2200 lbs]. Water is heavy stuff! Since clouds are composed of water, why don't they fall from the sky as soon as they form?

Air going up

The most common way to cool damp air enough to produce cloud droplets or ice crystals is to make it ascend. In some cases, a volume of ascending air may be a few hundred meters across; in others, it may be as much as a thousand kilometers across.

The speed with which air rises varies. In general, small cumulus clouds contain updraughts of 1–5 m/sec [3–16 ft/sec], whereas cumulonimbus clouds may have air rushing

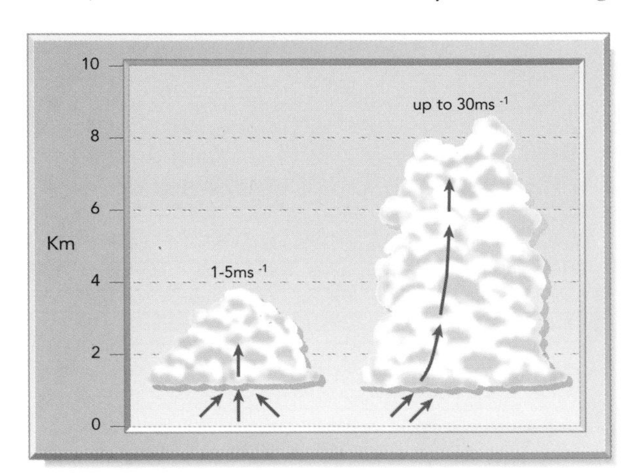

Updraughts in cumulus clouds

up at 30 m/sec [100 ft/sec] or more in severe cases! In contrast, extensive layer cloud, formed by widespread ascent, is associated with air rising at 5–20 cm/sec [2–8 inches/sec]. Here, in part, is the reason why clouds stay up: they are always associated with rising air.

Impure water

All cloud droplets have a nucleus around which they have condensed – known as the cloud condensation nucleus (CCN). The number of these varies from ocean to continent, and in height within the troposphere, but a typical value at sea level is around 100–200 million in every cubic meter!

Cloud particles

	diameter (microns)	number ($/m^3$)	fallspeed (m/sec)
Cloud condensation nucleus	0.2	100,000,000	0.00001
Cloud droplet (typical)	20.0	1,000,000	0.01
Cloud droplet (large)	100.0	1000	0.27
Small drizzledrop	200.0	1000	0.70
Raindrop (typical)	2000.0	1	6.50

The incredibly tiny cloud droplets are so small that their terminal fallspeeds are much lower than the speed of the updraughts that create the clouds. They settle at about 1 cm/sec [0.4 inch/sec], while the larger ones do so at 30 cm/sec [12 inches/sec]. Generally, the larger cloud droplets are found in convective clouds, where they can grow within the fast updraughts.

Falling drops

If, by some process, cloud drops grow large enough to attain a fallspeed greater than the ascending air speed, they will fall as rain. A borderline cloud/drizzledrop has a diameter of about 0.2 mm [0.007 of an inch], while a typical raindrop has a diameter of 2 mm [0.07 of an inch]. A raindrop of this size falls at up to 6.5 m/sec [21.3 ft/sec]. Drizzle is formed of drops with a diameter of 0.2–0.5 mm [0.007–0.01 of an inch]. Even if they do become this large and fall out of the cloud, they suffer some evaporation in the subcloud layer, between the cloudbase and the surface. If the air in this layer is dry and the raindrops small, they may actually completely evaporate on their way down.

Sometimes it is possible to observe this when a shaft of rain or snow can be seen falling from clouds. The shaft will narrow toward the surface and vanish above it. Such features are called virgae or fallstreaks. They can also be seen

Precipitation falling from cumulus clouds but not reaching the surface

as the trailing component of "Mares' tails" cirrus cloud, although the fact that they are precipitating ice crystals will not be obvious from the ground, since they exist quite a few kilometers up. This particular type of cloud occurs within a layer where the wind speed increases with height. The top part of the cloud is the source of the precipitating crystals, which fall through this zone to produce a trail that moves slowest at the base and quickest at the top.

Ascent in bubbles

Because pressure decreases with height in the atmosphere, any parcels (bubbles) or layers of air that ascend will expand gradually. This increase in volume causes the air within the rising bubble to cool.

Air not only rises, of course: it must also come down in places. In this event, the air sinks into steadily rising pressure. Under these conditions, the air is warmed.

So long as the ascending or descending air is unsaturated (cloud-free), it will cool or warm respectively at the rate of 9.8°C per kilometer [28.4°F per mile].

The birth of cumulus

Imagine a summer's day when the temperature is 25°C [77°F] and the dewpoint 15°C [59°F]. A bubble of cloud-free air pulls away from the heated surface as a thermal; the air is unsaturated so its temperature will decrease at a rate of 9.8°C per kilometer [28.4°F per mile]. If we assume that the concentration of water vapor remains constant as this happens – it can be increased if water is evaporated into the bubble and decreased if water condenses out of it – then after 100 m [328 m] the temperature would be 24°C [75°F], which has a lower saturation value for water vapor than at the air's surface temperature.

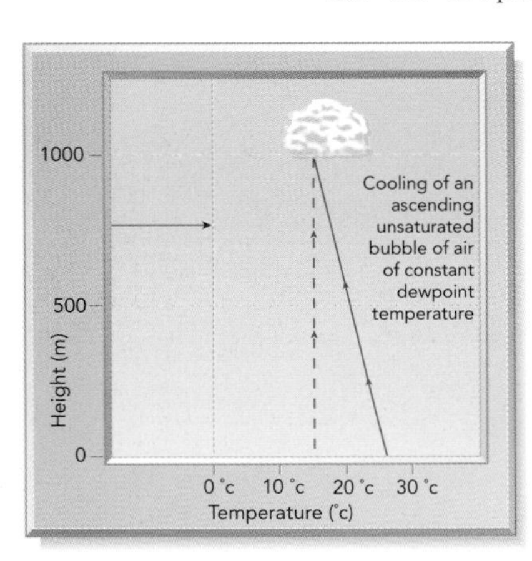

Expansion and cooling of an ascending dry air "bubble"

Continued ascent will lead to further cooling of the bubble and a lowering of the saturation value, until at 1000 m [3280 ft], the air will have cooled to its dewpoint temperature of 15°C [59°F]. If the air ascends beyond that point, the additional cooling will lead to the appearance of cloud droplets, formed from some of the water vapor in the bubble. This is where the cloudbase would be.

Once the condensation begins, a cloud appears. The type associated with the thermal described above would be a cumulus cloud. Condensation releases heat, which warms the surrounding air. This means that ascent within cloud causes the air to cool more slowly than in cloud-free ascent.

Impact of water

The scattered cumulus have a clear subcloud layer, where the unsaturated bubbles are ascending, and a cloudbase that marks the level at which they become saturated. The height of the cloudbase depends on the temperature and dewpoint of the surface air. If the air is dry, it needs to ascend by a considerable distance to produce condensation. If it is moist, the cloudbase will be nearer the surface.

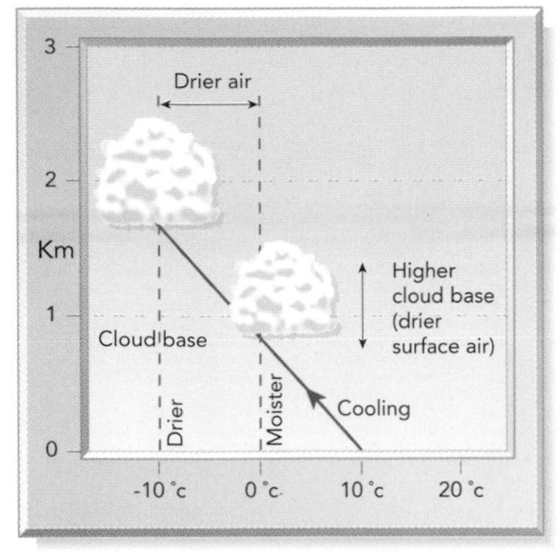

Influence of surface humidity on cumulus cloudbase

During the day, air just above the Sahara Desert ascends in response to the intense heating of summer. It rises as vigorous thermals to a substantial height above the desert, but generally is so dry that saturation is not achieved. This is known as dry convection.

The height a cumulus cloud grows to depends partly on the moisture content of the bubble – that is, how much heating is available from condensation and, therefore, how buoyant it will be. This is determined by the dewpoint temperature: the higher the value, the greater the absolute concentration of moisture. The thickness of the cloud is also affected by the way in which the temperature of the surrounding environment changes with height – known as the environmental lapse rate.

Growing cumulus cloud with cirrus above

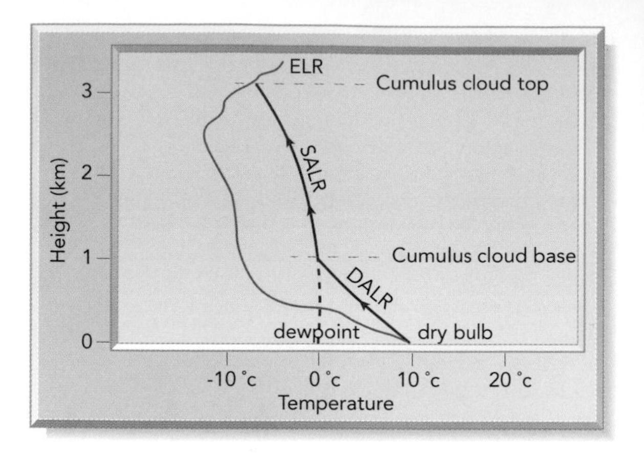

Parcel curve (red) for cumulus cloud developing within a cooler environment (blue)

DALR: Dry Adiabatic Lapse Rate
ELR: Environmental Lapse Rate
SALR: Saturated Adiabatic Lapse Rate

Impact of the environment

The environmental lapse rate is sensed by weather balloons as they ascend through the troposphere and lower stratosphere. To estimate the potential depth of a cumulus cloud, it is necessary to compare the environment temperature curve to that of the thermal.

At any one time, it is possible to observe cumulus clouds of varying depth scattered over areas a few hundred kilometers across. Such differences may be related to changes in the intensity of surface heating, or to the way in which the environmental lapse rate varies from place to place.

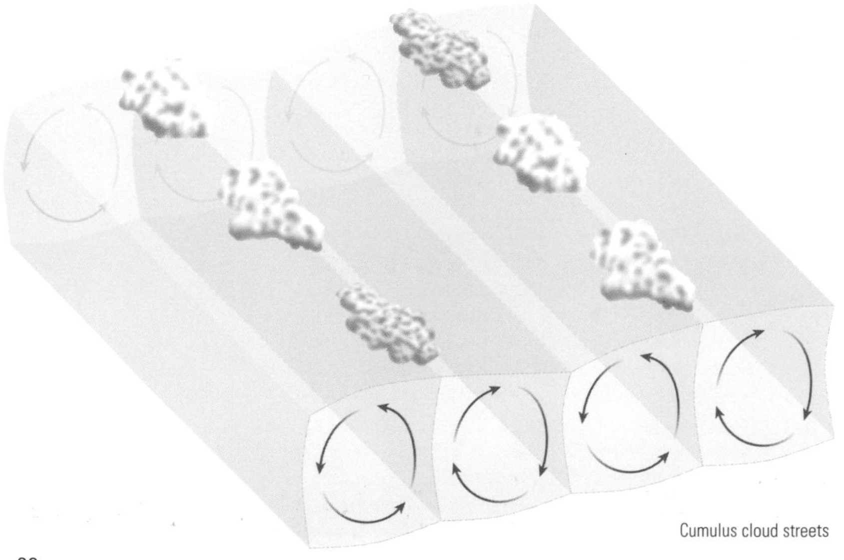

Cumulus cloud streets

Cloud streets

Some days, conditions can be such that convective clouds develop into a distinct pattern of long lines, separated by clear air. These lines can stretch for many tens or hundreds of kilometers along the direction of the wind and are known as cloud streets. They are formed when there is a temperature inversion (a rise in temperature with height) a few kilometers above the surface, and when the wind direction remains constant with height below this level, having a speed of at least 13 knots at the surface. The cloud streets align along the average direction of the wind and, typically, are spaced about 10 km [6 miles] apart.

Because the streets occur when an inversion is present, the clouds normally do not become deep enough to produce precipitation. They characterize pleasant, partly-sunny weather, and last until the larger-scale weather conditions that permit their development change. Cloud streets are quite common over mid- and high-latitude oceans in the autumn and winter, and over the land in spring and summer.

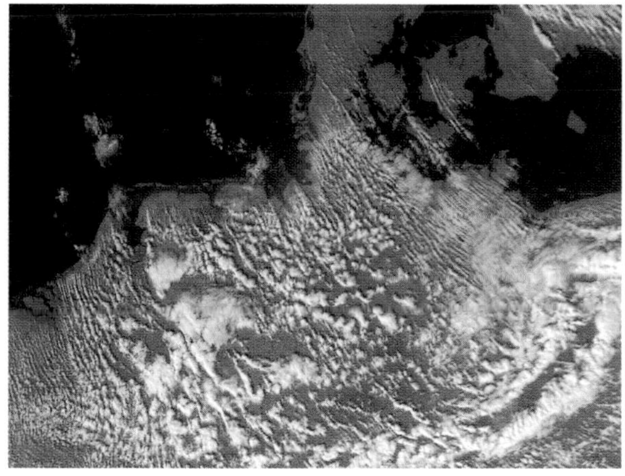

Cloud streets form Belgium to Poland on a March day

Cumulonimbus

At the other extreme of the cumulus scale lie the troposphere-deep cumulonimbus clouds, which are defined as low cloud because they have a base below 2000 m [6600 ft]. These are the tallest clouds we see, but because the troposphere's depth increases from pole to equator, they can reach up to 20 km [13 miles] in low latitudes, but only about 6 or 7 km [4 or 4.5 miles] in polar areas.

Because surface heating plays an important role in forcing the ascent of air from low levels, cumulonimbus clouds tend to occur more frequently across strongly-heated areas, above which relatively cooler and/or drier air occurs,

Cumulonimbus

the environmental lapse rate ensuring that bubbles of air ascend to great heights.

Unlike shallow cumulus, cumulonimbus clouds produce heavy rain and sometimes hail, thunder and lightning. As the cloud grows rapidly up through the troposphere, it is vital that the updraught, which forms the core of the ascending air, is separated from any precipitation that starts to develop within the cloud.

If this does not happen, any precipitation that forms will fall through the cloud-producing updraught, evaporating as it goes. This evaporating rain cools the air around it as it descends, which results in a cool sinking downdraught in the very core of the cloud.

This means that if a convective cloud grows in an environment where the wind speed and direction do not change with height, the cloud- and precipitation-producing ascent is "killed" once the raindrops fall back through it. Cumulus of this type commit suicide, producing only light showers in their short lifetime of a few tens of minutes.

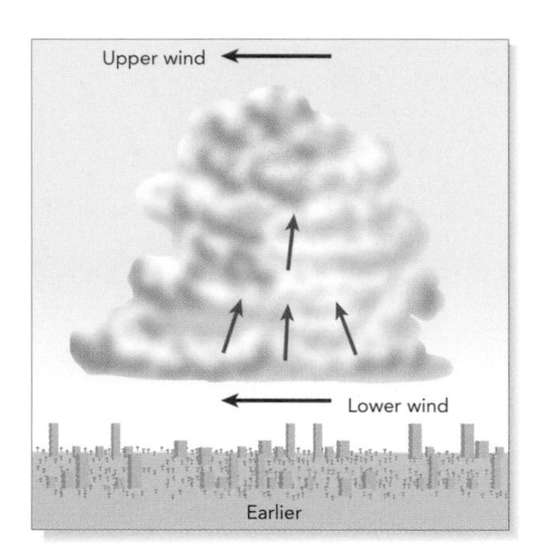

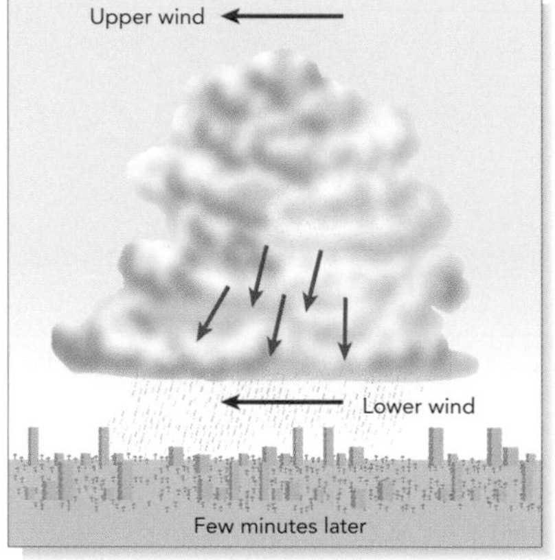

Short-lived cumulus in small/no wind change with height

If, however, the wind direction and speed do change with height, there is a chance that the updraught will be separated from the precipitation produced within the cloud. This is the recipe for a longer-lived, deep convective cloud that produces heavy, but generally short-lived, rain.

The region of the cloud where the rain occurs still experiences a downdraught produced by evaporative cooling, and this feature can be dangerous. When it reaches the surface, it spreads quite rapidly sideways as a cool, and often very hazardous, gust front. In hot, arid areas, a gust front and the air following it may be made obvious by becoming heavily dust-laden, suddenly cutting visibility to zero, as in the case of the haboobs of North Africa.

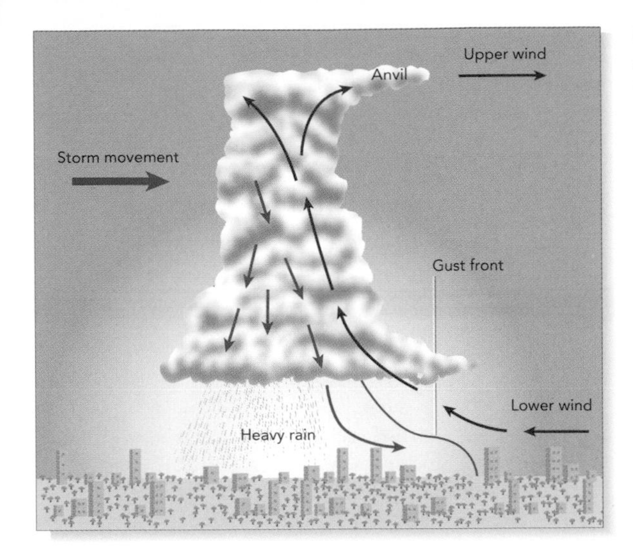

Longer-lived cumulonimbus in significant wind change with height

Gust fronts are particularly dangerous to aircraft landing and taking-off, because they can sweep across an airport some distance from the cloud in which they formed. Since aircraft take-off and land into the wind, any sudden change of wind speed and direction can have fatal consequences.

Arctic sea smoke

Occasionally, when cold air spills over much warmer water, the extreme temperature gradient through the air just above the water triggers very localized rapid ascent of bubbles of air, within which the water vapor condenses as narrow plumes. These features are known as Arctic sea smoke or steam fog. They can occur over open water in the Arctic and over lakes in middle latitudes in the winter.

Haboob or duststorm

Arctic sea smoke or "steam fog"

Ascent in layers

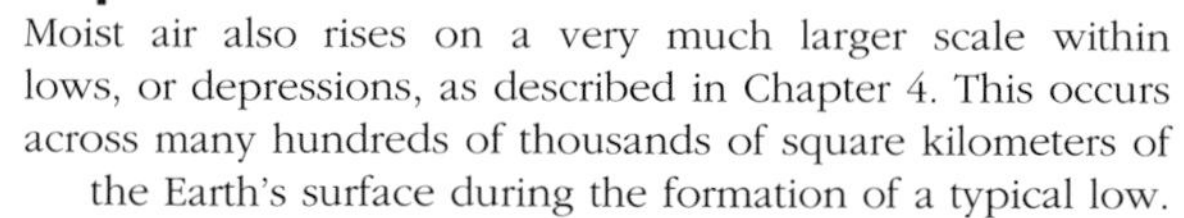

Cyclonic flow into a depression

Depressions

Moist air also rises on a very much larger scale within lows, or depressions, as described in Chapter 4. This occurs across many hundreds of thousands of square kilometers of the Earth's surface during the formation of a typical low. Therefore, depressions are cloud-laden, often with deep layer cloud that can produce widespread precipitation. As lows track across the Earth, the cloud is borne with them.

Some lows are frontal, which means that they have attendant warm and cold fronts, as well as an occluded front where the latter catches up with the former and lifts the warm air off the surface. Deep layers of warm, moist air ascend continuously over the gently-inclined warm and cold fronts, to produce very extensive condensation in the form of cloud.

We know that the winds circulating around a low, both at the surface and in the upper air, often travel at different speeds. This means that the air must be moving relative to the low as it travels across the Earth's surface.

Conveyor belts

In fact, frontal systems have characteristic large-scale currents of air that move in an organized fashion. The major cloud-producing flow is called the warm conveyor belt which streams through the warm sector ahead of the cold front. It ascends gradually from a kilometer or so above the surface, in the south-western part of the sector, as it runs parallel to the front, flowing over the warm front up to 5 or 6 km [about 40 miles] above the surface.

This feature is called a conveyor belt because it transports most of the all-important heat and moisture associated with frontal depressions. The very large thermal difference

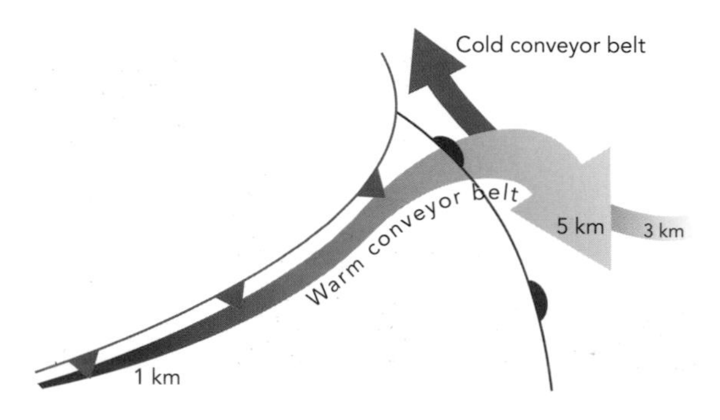

Large-scale flow relative to a middle latitude partly occluded depression and detail of precipitation pattern

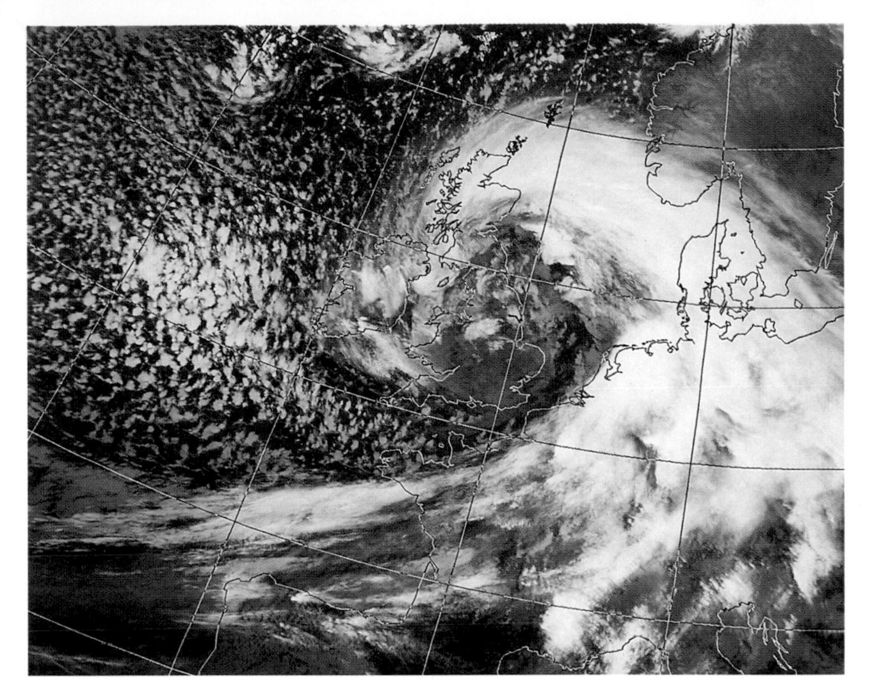

Frontal cloud and scattered showers of the Burns' Day Storm

between the tropics and extratropics drives the atmosphere and ocean to act in such a way that they propel warmer fluid polewards and cooler fluid equatorwards.

The warm conveyor belt is very closely related to the massive region of cloud within the warm sector and above the warm front. This cloud is the "signature" of huge volumes of tropical maritime air that flow polewards and upward within the frontal system. Both movements act to cool the air.

In addition, there is a cold conveyor belt, which consists of air that actually approaches the warm front from ahead and sinks to move parallel to the front, undercutting the higher warm conveyor belt. Then, it ascends into the occluded front. A third current flows through the middle troposphere, overrunning the traveling low as a cold, dry stream of air.

Behind the cold front, a huge volume of cool, relatively dry polar air streams across the surface. Over the ocean, in winter, a great deal of heat and moisture is pumped into the air as it flows across the generally warmer sea. Here, the depression is transporting cold, dry air equatorwards, and air that flows toward the equator tends to sink, becoming compressed. As a result, it is warmed by two processes: convection and compression warming of the air that sinks between the convective clouds.

This action is visible as a region of widespread, scattered convective cumulus clouds that rise essentially as bubbles from over the sea. They often produce showers, separated by bright spells, caused by the sinking air.

Hills and mountains

Air has no choice but to flow over and around upland areas. If the air is damp, its forced ascent (termed orographic uplift) can often lead to saturation and condensation into orographic cloud. The amount of uplift required depends on how close the air is to saturation point. Very dry air will require a good deal of cooling (uplift) to reach its dewpoint temperature, while very damp air will need only the slightest ascent to produce cloud.

Wetter hills

It is within warm sectors that rain is commonly subjected to a process known as orographic enhancement. Upland regions of South Wales in the UK, for example, can experience a fall of rain that is two or three times more intense than that falling on the coast at the same time. Careful study of this phenomenon has shown that key ingredients are required to produce significantly heavier rain in upland areas. It is not true that simply because a moist airstream crosses a hilly district, it will generate more rain.

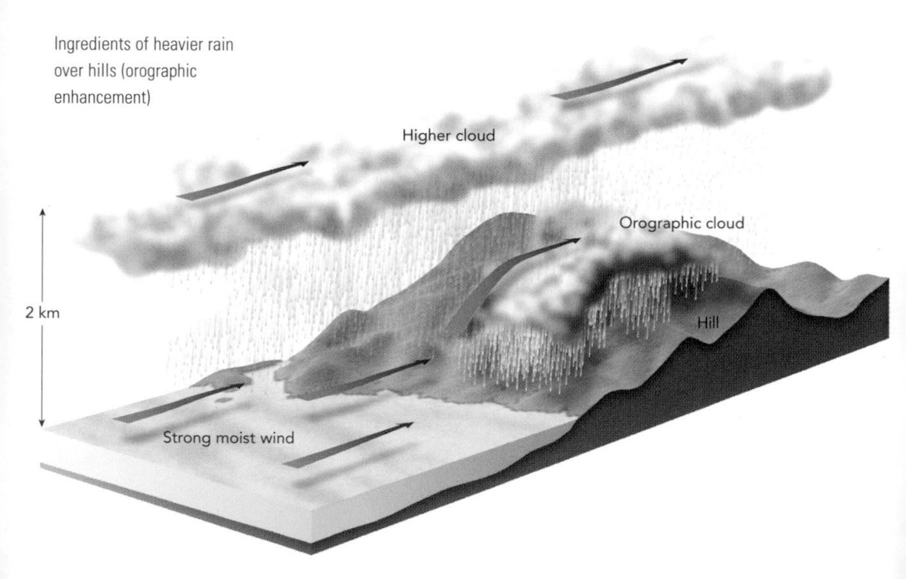

Ingredients of heavier rain over hills (orographic enhancement)

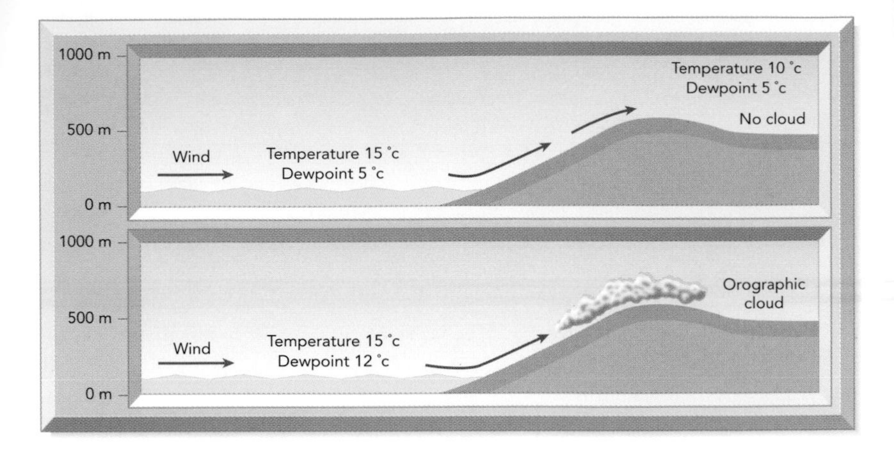

Role of low-level humidity on the presence of orographic cloud

Orographic cloud is quite common when condensation develops in moist air that has cooled because of its forced ascent. Therefore, hilly areas are often cloudier than adjacent lower land.

Orographic enhancement occurs in warm sectors when there is a precipitating layer of cloud at a height of about 2–3 km [1–2 miles]. This layer will not be related to the hills in any way, but to the large-scale flow of the frontal depression. Sometimes, as this layer moves across the hills, the rain it produces falls through cloud that has been generated by a strong, low-level stream of moist air flowing up and over the hills, made visible by orographic cloud.

While this situation persists, the rain washes out large quantities of water from the lower cloud, thus increasing rainfall over the hills compared to other areas. If the lower cloud is constantly replenished by a strong surface flow of damp air, there will be a prolonged period (over many hours) of orographic enhancement. However, if the surface flow is weak, the water will soon be washed from the cloud, and not replenished at a rapid enough rate, providing only a fleeting addition to the catch over the hills.

This action occurs across many middle-latitude hilly areas that are frequented by frontal depressions. It is the explanation behind the wet reputation of upland North Wales, Western Scotland and the Lake District in the UK.

Lee waves

Under special conditions of wind and temperature change with height, hills and mountains can generate standing waves in the airflow above and downwind of them. These lee waves are indicated by cloud in the ascending air, and by clear areas where the air descends. Such waves can appear over a distance of 100 km [63 miles] or more, to

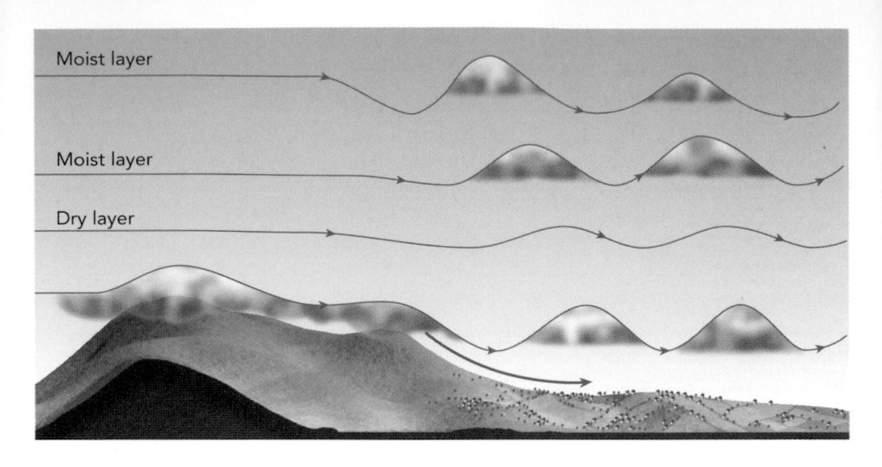

Generation of lee wave clouds

leeward of upland areas. They are called standing waves because they are anchored to the hills or mountains that produce them.

Lee waves can be compared to the standing waves often seen in streams where water flows over large stones. The flow creates a pattern downstream, in which the waves are stationary, but the water flows quickly through the pattern.

The ribbed cloud pattern associated with lee waves will exist for some hours, rather than days, until the large-scale weather pattern that favours their development changes. The lenticular (lens-like) clouds produced by the wave pattern do not produce precipitation.

On some occasions, the droplets that compose the lee wave clouds freeze and, if the air is humid enough, a very extensive sheet of ice crystals develops, which will be carried many kilometers downstream by the wind. The true extent of such orographic cirrus can be seen from satellite photographs.

Fog

The official definition of fog is a condition where the horizontal visibility is 1000 m [3280 ft] or less because of the presence of water droplets suspended in the atmosphere. Thick fog occurs when the visibility is 100 m [328 ft] or less. Impaired visibility of more than 1000 m is defined as mist.

Inland on a cloudless, calm night when the air has low humidity, there will be a large flow of radiation from the Earth's surface and atmosphere out into space. If cloud is present, its water vapor, water droplets and ice crystals will absorb some of this outgoing energy and radiate some of it back down to the surface and lower levels of the atmosphere. Therefore, cloud – especially layer cloud – acts like an insulator.

Lee wave clouds over Italy, Albania, Macedonia and Kosovo (north-south aligned "ribbed" clouds)

Orographic higher level cloud streaming away north-eastwards from the Pyrenees

If the sky is cloud-free, but the air is humid, the water vapor present will also absorb some of the outgoing radiation and, like a cloud layer, will radiate some of it back to the surface and lower layers, keeping them warmer than they would be otherwise. If, however, the air has very low humidity, much of the heat will escape to space, and the surface will be much chillier. One way of determining whether cloud-free air is relatively dry or moist is to watch the stars. If they twinkle, humidity in the air is refracting the light coming from them; if they do not twinkle, the troposphere is quite dry.

Once the Sun is low in the sky and the air begins to cool, it does so most strongly at the surface. Thus, the chilling tends to be most marked at and near the ground, notably on calm nights. This means that a temperature inversion will form above the surface, where the temperature increases with height. If the conditions are calm, or near calm, the air adjacent to the surface may cool until it reaches its dewpoint. Light, subtle movement will spread the cooling through the surface layer; any stronger motion – if the wind picks up – will mix the warmer air above and the chilly layer below, destroying the conditions that favour fog formation.

Radiation fog

Calm, cloud-free conditions can produce radiation fog. The word "radiation" expresses the means by which the air is

Surface chilling at night in dry and humid atmospheres

In very cold conditions, in a very dry atmosphere, there is a strong cooling to space

In cold/cool conditions, when there is some water vapor, but it is cloud-free, the cooling to space is offset by re-radiation downward

cooled to its dewpoint temperature – the Earth and the atmosphere lose heat rapidly by radiating it to space.

Radiation fog in the valley around Neuschwanstein Castle, Germany

Once the fog develops and grows vertically, the effective radiating surface is no longer the ground, but the top of the fog. The temperature inversion is found at the top, too, often many meters above the ground.

Radiation fog is most common when chilling is strongest, during the autumn and winter, and it is confined to land areas. Its frequency depends on the distance from the sea and the local lie of the land. Such fog tends to occur more frequently across low-lying areas, like valleys, into which cool air drains slowly during the hours of darkness.

The sea cools only marginally at night – considerably less than the land surface does. In fact, marine cooling is so minimal that it does not lead to radiation fog.

Evolution of radiation fog during the night

At sunset, in clear, calm conditions, radiative cooling occurs up through damp air

A few hours later, radiative cooling continues and fog occurs at the surface

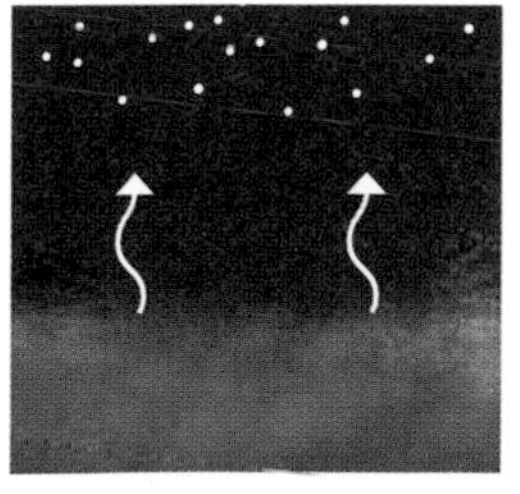

Later still, the fog deepens, as the radiative cooling continues under the clear conditions

Advection fog

Cooling of the air can also occur when a warm air mass flows across a colder surface, in which case heat is transferred downward from the air. This can reduce the air temperature to its dewpoint, producing saturation, then fog. The critical difference between advection fog and radiation fog, is the role played by air movement in its formation. The term "advection" is used almost exclusively in meteorology and oceanography, normally referring to horizontal motion that transports some property of the fluid. For example, "thermal advection" refers to the amount of heat transported by the wind or ocean currents.

Advection fog is most commonly found in areas of poleward-moving tropical maritime air that is cooled by contact with the sea's surface. Thus, it is also known as sea fog. It occurs most often in the spring and early summer, when the sea's surface temperature is at, or recovering from its lowest. In the seas around Britain, for example, especially to the south-west from where the tropical maritime air most often approaches, advection fog is quite common. On Britain's east coast, too, the cooling of moist onshore flow leads to the development of the "fret" along the Northumbrian coast, and "haar" across stretches of south-east Scotland. These are regional names for the same phenomenon.

Advection fog lapping the Californian coast

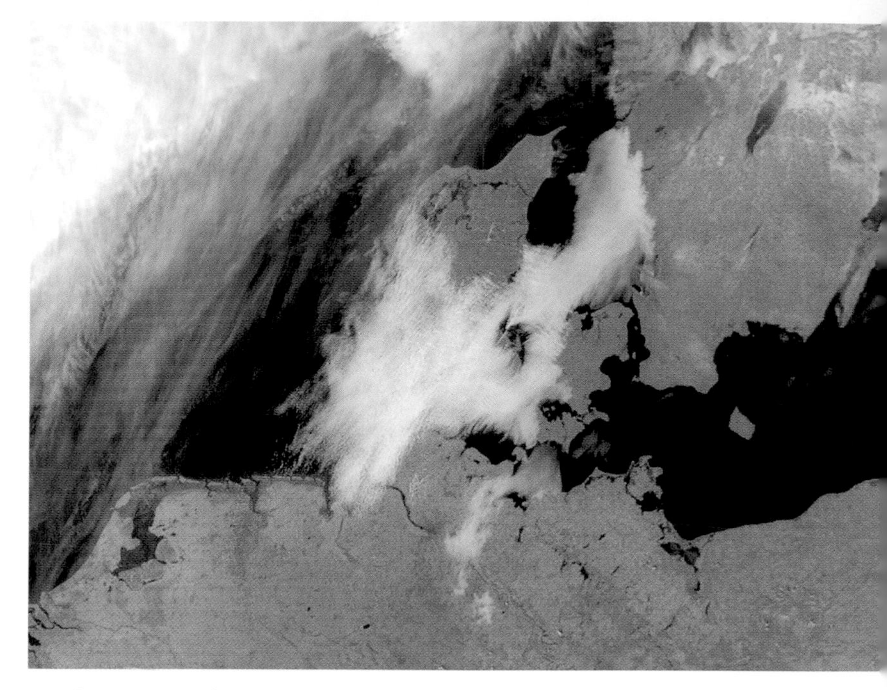

Advection fog across the waters between Denmark and Sweden

Other persistently cool areas of ocean witness much more frequent advection fog, although it is not very common within the tropics. Among these regions are the Grand Banks, off Newfoundland, where, in July, advection fog occurs on four out of ten days over the cool waters of the Labrador current. It is as common over the cool Oya Shio and Kamchatka waters in the north-west Pacific, as it is in the Bering Strait. In higher latitudes, sea fog is frequently found over the pack ice and open waters of the summertime Arctic Ocean and Canadian archipelago, and to some extent over the pack ice and open waters around Antarctica.

Coastal advection fog often occurs where unusually cold sea water flows parallel to subtropical western continents. Strong cooling of the low-level air leads to fog along the coast of northwest Africa (over the Canaries Current), south-west Africa (the Benguela Current), Chile (the Humboldt Current) and, perhaps most famously, the central and northern California coast.

By definition, advection fog moves. This means that even with winds of 30 knots over the sea, thick fog may still be present. However, with strengthening wind, the fog often lifts to form extensive stratiform cloud.

Although advection fog is most common over the sea, it can occur over land when warm, moist air passes across a snow-covered surface or one that has recently been frosty.

Distant advection fog over patchy snow in the Thames Valley, England

Hill fog

Another frequent type of fog is hill fog, which occurs when layer cloud intersects a range of hills, reducing visibility in those portions of the hills within the cloud to 1 km [0.6 of a mile] or less. Hill fog often occurs in moist warm sectors of frontal depressions, where the cloud base is low.

Dew

Dew forms by the direct condensation of water vapor on to the ground, most noticeably on grass. Dew will occur under conditions that favour the generation of radiation fog. It is deposited before such fog develops, but is often observed when there is no fog at all. On these occasions, the cooling is sufficient to produce a dewfall, but is not intense enough to affect condensation within the lowest layers of the atmosphere. In regions where precipitation is very sparse, dew can provide an important source of water, both for plants and animals.

Frost

The most common form of frost is hoar frost. It is the equivalent of dew, but the water vapor is deposited as ice crystals in the form of scales, needles, feathers, etc. on blades of grass, bushes, and other surfaces. Like dew, hoar frost develops under clear, calm conditions. The temperature to which the air must cool to produce frost is not the dewpoint, but the frost-point. This is defined as the temperature to which the air must be cooled (at fixed pressure) to saturate it with respect to an ice surface, rather than a liquid water surface. Hoar frost is the typical "Jack Frost on the window pane" of many a home in the winter before the advent of double glazing and central heating.

A less common form of frost, which often produces quite dramatic forms, is rime. This occurs when supercooled cloud and fog droplets come into contact with cold surfaces to form masses of white ice crystals. Rime is most commonly encountered in upland areas during winter. Sometimes, quite amazing shapes may be observed because the crystals are deposited while the supercooled cloud or hill fog is in motion. The frost formation grows downstream of the object on to which the deposit was first made.

It is worth noting that visible frost does not always occur when the air temperature falls below 0°C [32°F]. Sometimes, for example, when the British Isles and other parts of Northern Europe are in the grip of very cold easterlies in the winter, the air is so dry that overnight chilling is not intense enough to squeeze any water out of the air as a deposit of frost. Nevertheless, if the surface temperature reaches or falls below 0°C [32°F], ground frost is reported.

Hill fog

Thick hoar frost deposited on leaves and exposed roots

Rime deposit, Snowdon, Wales

Precipitation

For precipitation to occur, there must be a means by which cloud droplets or ice crystals can grow larger and heavy enough to fall as drizzle, rain, snow, or hail.

One thing we are all familiar with is the fact that water freezes at 0°C [32°F]. This is true under normal conditions, but in the atmosphere, where water particles exist as extremely small cloud droplets, this is far from the case.

Supercooled water

Even at high altitudes within the troposphere, many cloud particles remain liquid in what is termed a supercooled state. Except at very low temperatures, liquid water will not freeze unless minute impurities are present. These are much less likely to occur in the very small droplets that form clouds than in substantial bodies of water where freezing occurs at 0°C. Therefore, it is better to define 0°C as the melting point of ice.

In the atmosphere, only one cloud droplet in a million is frozen at –10°C [14°F], a couple of hundred or so in a million are frozen at –30°C [–22°F], and only at –40°C [–40°F] and below will they all be ice crystals. The most effective surface to freeze water upon is an ice crystal, so if supercooled droplets touch one, they freeze instantly. This means that in clouds where both supercooled droplets and ice crystals are present, the crystals grow rapidly.

Plate

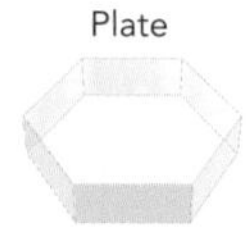

Column

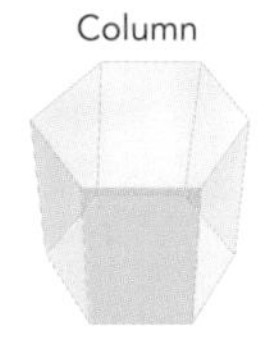

Dendrite

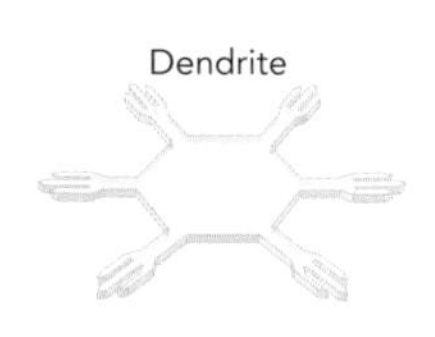

Common forms of ice crystal

Ice crystals

Ice crystals that grow from vapor alone take on characteristic shapes depending on the temperature range within which they are created. As they descend through progressively warmer layers, they become more complex in shape. Similarly complex form changes can occur if they ascend on updraughts into cooler regions of a cloud. The table below summarizes the forms of these crystals:

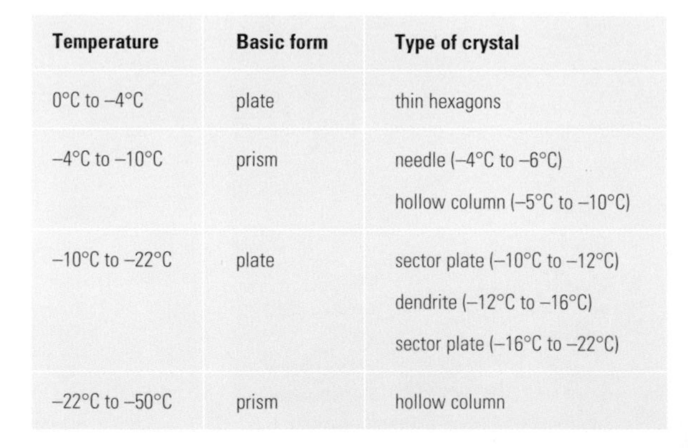

Temperature	Basic form	Type of crystal
0°C to –4°C	plate	thin hexagons
–4°C to –10°C	prism	needle (–4°C to –6°C) hollow column (–5°C to –10°C)
–10°C to –22°C	plate	sector plate (–10°C to –12°C) dendrite (–12°C to –16°C) sector plate (–16°C to –22°C)
–22°C to –50°C	prism	hollow column

The freezing of supercooled water on to ice crystals is a second mechanism of growth. It is known as riming, which is essentially the same process that causes the deposit of rime as a frost. Crystals may grow at varying rates depending on how much supercooled water freezes on to them, larger ones can capture others as they fall at higher speeds.

Crystals into flakes

The icy particles formed in this manner are known as graupel, which fall and fracture when they crash into cloud droplets. Such splinters can grow into new graupel, which may fragment again to produce a chain reaction, forming very large numbers of ice crystals. As these descend, they often stick together to produce snowflakes. In fact, most of the rain we see in middle and higher latitudes starts life as snow, even in the summertime.

Dendrite snow crystal

Warm clouds

In contrast to the "cold" clouds that produce ice crystals, others with tops warmer than –15°C [5°F] often generate precipitation in an altogether different way. These "warm" clouds are composed of cloud droplets of varying size, which collide as they settle. The larger drops fall faster and sweep up smaller drops by a process known as coalescence. The number of raindrops that form within these clouds depends on the liquid water content, the range of droplet sizes, the strength of the updraught within the cloud – which determines the time available for a droplet to grow – and even the electrical charge carried by the droplets.

Thus, drizzle falls from shallow stratus, within which rather weak upcurrents of some 10 cm/sec [4 inches/sec] occur, while very vigorous tropical cumulus cloud will generate updraughts of many meters a second to produce raindrops of up to 5 mm [0.2 of an inch] in diameter.

Rainbows

Sometimes, when the weather is showery, we may see rainbows. These are visible when the Sun shines upon the falling drops – and it must be shining from behind us as we look toward the shower. This means that in the morning, rainbows will be visible in the west, and in the afternoon, broadly in the east.

When sunlight enters a raindrop, some of it passes straight through, while the remainder is reflected back by the rear surface of the drop. The angle at which this occurs is about 42 degrees; as the light enters the raindrop, each ray is refracted slightly differently, as is each ray leaving the drop. When combined with the internal reflection, this double refraction splits the "white" sunlight that shines on to a drop

Primary and secondary rainbows

Primary and secondary rainbows

into its component colors, in the same way that a prism splits white light into the colors of the spectrum.

When this happens within a mass of falling raindrops, we see a rainbow. Refracted red light enters our eyes from higher drops, and violet light from lower drops. As a result, the brilliant rainbow we see is red at the top and violet at the bottom. Occasionally, there may be a fainter, but noticeable, secondary rainbow. This forms when sunlight enters the raindrops at such an angle that a double internal reflection occurs. As a result, the light that finally leaves such drops is a lot fainter and the colors weaker.

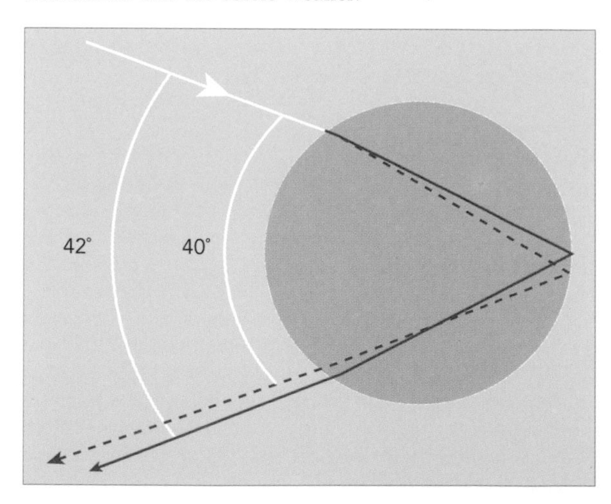

Refraction and internal reflection of white light shining on to a raindrop

chapter six

6

Forecasting the Weather

To most people, the main aim of the meteorologist is to prepare and present a weather forecast. Furthermore, television is probably the most common medium through which they obtain information on what is likely to happen over the next day or so weatherwise.

Weather lore

It is not really surprising that the kinds of traditional observations, which refer to the nature of the forthcoming season, rather than that of the next day's weather, are unreliable. Sayings that link the coming season to the amount of berries on particular trees, for example, are gross oversimplifications. Despite this, they are often interesting, and occasionally traditional sayings relating to short-term forecasts contain a grain of scientific truth.

Season ahead

The challenge of forecasting the nature of a forthcoming month or season is immense and is being pursued scientifically today. It is not the case that the character of, for example, a summer – whether it will be warm and dry or cool and wet – is necessarily strongly related to the preceeding spring or winter. A glut of berries is an expression of the quality of the season during which they grew. They are not reliable harbingers of subsequent months.

The notion that the weather on a particular day of the year may indicate the nature of the weather to come over the following month, or more, is the basis of Groundhog Day in the USA, and St Swithun's Day in the UK.

Groundhog Day, on 2 February, takes place at Punxatawney, Pennsylvania, and involves a captive groundhog known as Punxatawney Phil. If he sees his shadow when he pokes his head out of his burrow, winter will last another six weeks; if he does not see it, spring is just around the corner. This tradition was apparently imported into the USA from Europe.

Red sunset across Morecambe Bay, England

If the day is sunny and, therefore, probably cold and anticyclonic, local folklore has it that these conditions will persist. If there is no shadow, it is obviously cloudy and probably milder. However, there is no compelling evidence that this particular day is the key to forecasting long-term weather patterns.

The same is true of St Swithun's Day on 15 July. Swithun died in 862, when he was Bishop of Winchester, and was buried at his own request in the cathedral grounds. During the next century, he was canonized and the decision was taken to move his remains to the choir of the cathedral on 15 July. The plan was abandoned, however, after 40 days of rain that began on that day and, in due course, the event passed into folklore. In the cold light of scientific reality, a wet St Swithun's Day is extremely unlikely to start a run of wet weather lasting nearly six weeks in Winchester or anywhere else.

Day ahead

Over shorter time scales, careful observation of indicators provided by the atmosphere can lead to a successful prediction. This type of weather lore is based on intelligent assessment of the signs in the sky, and how certain distinct features or changes can lead to specific events. None is foolproof, however!

The classic "red sky at night, shepherd's delight; red sky in the morning, shepherd's warning" works some of the time, but only in regions where the weather systems arrive from the west. The rule is based on the fact that redness during times of low Sun means that the weather in that particular direction (east in the morning; west in the evening) is settled. If the eastern sky is red, disturbed conditions will probably approach from the west. On the other hand, if the western, evening sky is red, the disturbed weather will have passed over to the east and the outlook will be for more settled conditions.

The morning warning does not apply during anticyclonic conditions when high concentrations of dust particles lie trapped beneath the inversion that characterizes a high. In this situation, both morning and evening experience red skies.

There are literally hundreds of weather sayings, some of which apply to specific geographical regions, while others are widely applicable. Nowadays, we are much less weather sensitive than our ancestors, whose careful observation of the sky produced the well-known lore. Their livelihoods may have depended on the following day's conditions.

Nevertheless, even today, we can learn from the changes in the character of the sky, especially the way in which clouds evolve over periods of minutes or even hours.

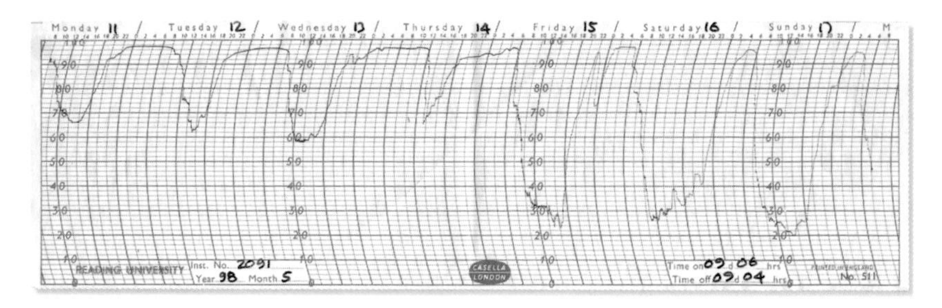

Change of relative humidity during one week with night-time maximum and daytime minimum

The lore relating to changes in seaweed and pine cones is also worth examining, although it is not particularly reliable. This rule suggests that changes in the bulk or shape of seaweed or pine cones indicate the likelihood of rain. It is based on the response of plant life to variations in relative humidity. When this increases with the approach of frontal rain, seaweed and pine cones will absorb some of the moisture which changes their appearance. Seaweed becomes more plump or less dry, and pine cones will partly close. Such changes are humidity-related, but may simply reflect the daily variation in relative humidity under settled conditions, which occurs as the air temperature increases to a peak in the afternoon, then decreases to a minimum during the night. There will be changes on the approach of moister air ahead of a depression, but for the vast majority of us, the sky is a far more reliable source of harbingers of a frontal system.

Forecasting today

The meteorologist who appears in front of millions of viewers is at the sharp end of the forecasting process. What he or she says about the weather represents the culmination of a truly global effort to collect, transmit, and process vast amounts of weather data from a variety of widely-scattered sources. The television forecaster is backed by a cast of thousands around the globe, whose tasks may include releasing radiosondes in the Antarctic, piloting large commercial aircraft, or serving on coastguard duty. The list of full- and part-time providers of weather data is virtually endless. Not only do national weather services maintain observation sites, but the wide range of professionals whose jobs are affected by the vagaries of the weather are stalwart providers too.

Data exchange

The truly international, free exchange of weather data is a hallmark of the profession – and necessarily so. The atmosphere knows no national frontiers; it is only in times of international belligerence that such information is withheld to disadvantage an adversary. So, what you see on television, hear on the radio or read in a newspaper is unique; no other profession provides forecasts with such frequency and under such scrutiny. To add to the pressure, many people believe that they are their own weather experts! Unlike a host of other technical experts, the meteorologist has to deal with data that is tangible to everyone.

How are forecasts made?

The raw ingredients of a weather forecast are the observations taken simultaneously around the world from widely varying platforms. One of the world's leading weather forecast centers is the European Center for Medium-Range Weather Forecasts (ECMWF) in Reading, Berkshire, in the UK. The data it collects on a typical day includes:

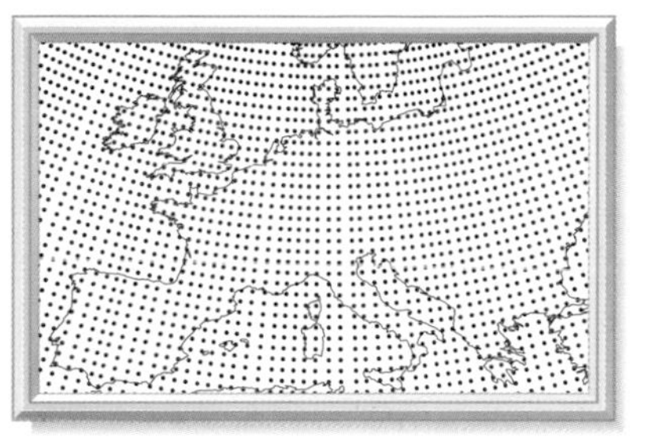

Grid points across a sector of the ECMWF operational global forecast model

8583 SATOBs, which are estimates of the wind speed and direction gained by tracking clouds from geosynchronous satellites like Meteosat. Only clouds that are known to be reasonable tracers of the air's motion are used. They are monitored within a circle of radius 55 degrees of latitude centred on the sub-satellite point, mainly for upper and lower tropospheric levels;

137,790 SATEM and TOVS (TIROS Operational Vertical Sounder) vertical profiles of temperature and humidity, sensed by NOAA polar orbiting satellites. The former are reduced-resolution versions of the latter, which are received direct from the satellite. SATEMs are transmitted via the WMO's Global Telecommunications System after processing in the USA;

1,054,938 observations from ERS-2 (the European Remote-Sensing Satellite) giving estimates of sea-surface winds. These are deduced from the satellite's measurement of the roughness of the sea's surface using radar;

32,710 AIREP, AMDAR (Automatic Meteorological Data Relay) and ACAR aerial observations from commercial aircraft, either automatically sensed and transmitted, or sent directly from the flight deck by senior staff;

European Center for Medium-Range Weather Forecasts, UK
http://www.ecmwf.int

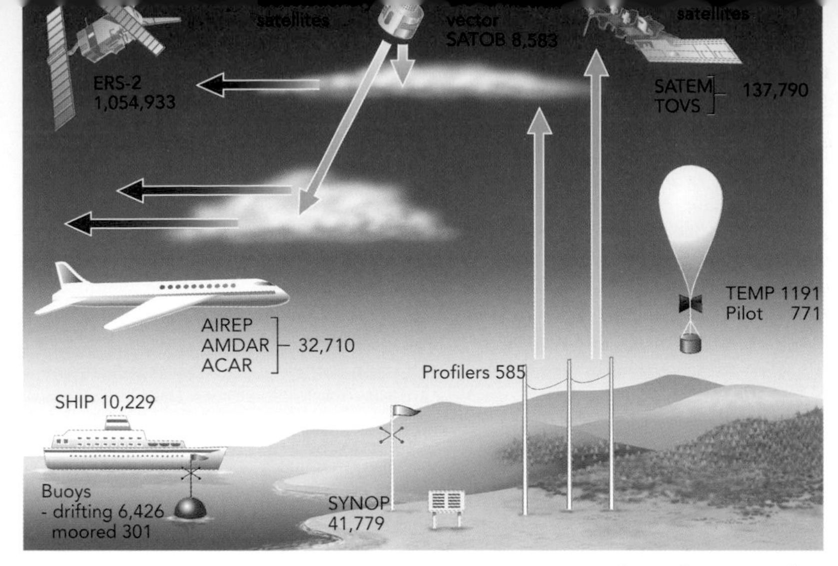

Numbers of observations received on a typical day at ECMWF, England

1191 TEMP reports representing the information from radiosonde ascents made around the world. They include data on temperature, humidity, and wind direction and speed up to a height of about 30 km [19 miles];

771 PILOT observations taken by tracking optically the drift of small balloons to deduce the winds, mainly in the lower troposphere. The data are restricted by the balloon vanishing into cloud or becoming too small to see;

585 Profiler observations from automatic instruments that sense wind strength and direction in the lower troposphere. Currently, these are restricted to the USA;

41,779 SYNOP weather reports from the traditional surface weather stations on land, including the growing number of automated sites;

10,229 SHIP surface observations, mainly from roving commercial vessels;

6727 observations from moored and drifting buoys, which complete the surface picture across the oceans.

To be able to predict what will happen to the weather over the next few hours or next few days, or even a week or more ahead, it is essential to gauge what is happening *now*. Observations that are taken around the world at 0000 and 1200 UTC are the data most often used to provide the ingredients for the computer models that form the basis of predictions.

Meteorological Office, UK
http://www.meto.govt.uk

Computer models

Meteorological computer models apply the unevenly-scattered surface and upper-air data to a regular latitude/longitude grid that has a number of levels up through the atmosphere. At ECMWF, for example, the operational global forecast model grid currently has a horizontal spacing of about 60 km [38 miles]. At each of the 138,346 grid points on the Earth's surface, there are values of temperature, wind direction and speed, and humidity, plus soil moisture and snow cover. The Center's operational model has 31 levels, in the vertical, too. These stretch from the surface up to 30 km [19 miles]. This means that calculations of the future values of temperature, wind, and humidity are currently carried out at 4,154,868 points through the atmosphere.

Some of the raw data cannot be made available when the main observations are taken from satellites, for example. These asynoptic data can be accepted into the cleverer computer forecast models when they arrive – as the models are actually running – to produce a prediction.

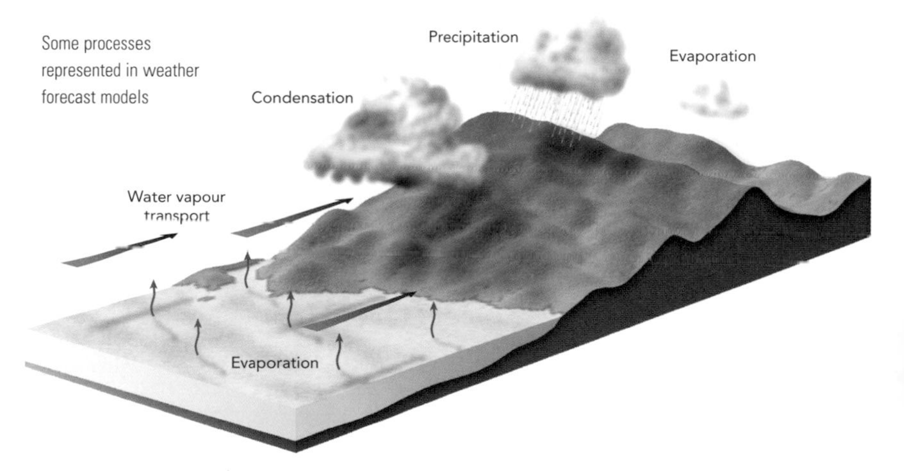

Some processes represented in weather forecast models

Predictive equations

The computer inserts the values of temperature, humidity, wind and so on into special predictive equations that forecast how the situation will change over short periods of time, like a few tens of minutes, at every grid point. Many more values are assessed at these points, including the all-important vertical motion of the air's flow. The forecasts for 12, 24, 48 hours, etc., ahead are generated by running the model over these short time steps out to 10 days ahead.

Computer models must also have information on how the intensity of solar radiation varies during the day and

National Weather Service, USA
http://www.nws.noaa.gov

throughout the year, and how the clouds, dust, and other particles in the atmosphere absorb, re-radiate, and reflect the radiation on its way to the surface. Therefore, they must generate clouds by evaporating water from the surface, they must transport them on their model winds, and condense some of the vapor into cloud water and precipitation – either widespread and perhaps prolonged, or short and sharp over localized areas.

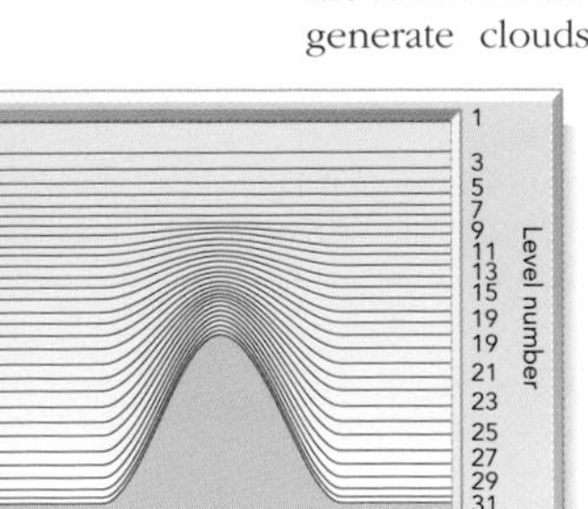

Levels in the ECMWF operational global forecast model

The passage of outgoing terrestrial radiation must be simulated, along with the quantity that is radiated back to the Earth's surface. The variations in vegetation, sea-surface temperature, and soil moisture should all be represented in these models. Overall, they are amazingly complex, but continue to improve as our knowledge of the workings and interactions between the atmosphere, the hydrosphere, the lithosphere, the biosphere, and the cryosphere advances.

Chaos

A notion that has thrown a spanner in the works of weather forecasting in recent years is that of chaos. What this boils down to is that a small change in the initial conditions of a forecast can lead to a dramatically different prediction. Initial conditions are those used to drive the weather forecast – that is, the data received from around the world from the 0000 or 1200 UTC observations.

The classic example of a butterfly flapping its wings on one side of the globe, and creating a storm on the other, is a way of saying that a very small input to the system – in a region that is sensitive – can lead to a significant, larger effect elsewhere. Quite whether the flapping butterfly will have any influence depends on when and where the flapping goes on.

Ensemble forecasts

It is impossible for operational meteorologists to represent such a small-scale influence but what they can do is run more than one prediction. This is known as the ensemble method, where many forecasts are run, using slightly different initial conditions for each. The marginal differences simulate the existence of "flapping butterfly wings", albeit on a rather larger scale, in various locations around the world.

At ECMWF, each operational ensemble run currently involves a total of 51 subtly different initial conditions. Of necessity, these forecasts are at a lower resolution than the main operational forecast, so a grid spacing of 120 km [about 75 miles] is used.

Once the forecasts have been produced, they are grouped into ensembles that represent broadly similar predictions. On some days, for example, there may be 30 that point in the same direction, with 12 indicating a different prediction. On another day, the clusters of similar forecasts may be many and consist of only three or four each. In this way, the forecasters can gage the likely accuracy of the prediction for a certain day. Sometimes, many forecasts indicate the same train of weather events, while on others they are widely different.

Benefiting from this massive computing power and model sophistication is the forecasting team. They have the task of interpreting and communicating the model products to an audience. This role has changed enormously over the last decade or so, and will continue to change to keep pace with modern technology.

D-I-Y forecasts

As we have already seen, weather lore occasionally works, but often, it fails to produce accurate forecasts. An improved method for assessing the weather is to study the way in which clouds evolve over a range of time scales.

Showers, sunny spells, or overcast?

Imagine a day when the morning sky starts clear blue, but pretty quickly small scattered cumulus clouds appear. These signify moist bubbles of air that have reached their condensation level. The amount by which they grow up into the troposphere is crucially important in determining the nature of the weather for the day that follows. Simply noting how cloud formations change over tens of minutes, or a few hours, can lead to an effective forecast for the day ahead.

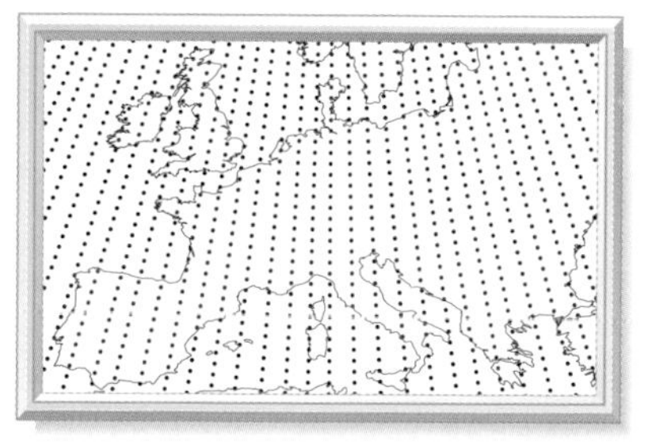

Coarser grid point spacing across a sector of the ECMWF global Ensemble Prediction System

If the troposphere is unstable – that is, the temperature of the environment through which the cloud is ascending is falling more rapidly than within the cloud – the cumulus cloud will grow upward with classic "boiling" upper reaches. On a good day, it is possible to watch such a cloud for a few minutes and actually observe its top evolving. These "cauliflower" cumulus clouds are known as cumulus congestus and indicate that the atmosphere is very unstable – where deep, moist convection penetrates successfully to great heights. If the clouds reach this stage, there is a good chance that a shower will fall from them.

Short- or long-lived

Such shower clouds last for several tens of minutes and produce a swathe of precipitation that may be as wide as the cloud, and up to 10 km [6 miles] or more in length. Visibility may be poor in the precipitation, and conditions can be gusty due to the precipitation-induced downdraughts.

Deep cumulus cloud

If the convection reaches the top of the troposphere, thunder is likely. A crucial aspect that determines whether such clouds become longer-lasting cumulonimbus or shorter-lived, moderately-deep cumulus clouds is wind shear.

If the convective cloud grows in a layer within which the wind does not change much with height, the precipitation that falls back through the cloud's updraught tends to kill it. If, however, there is a marked increase in wind speed with height, the cloud- and precipitation-producing updraught becomes separated from the precipitation-induced downdraught. This separation is a vital ingredient in the longevity of these much deeper convective clouds. Ultimately, they are limited in growth by the stable "lid" of the tropopause – the zone that caps the troposphere within which temperature remains the same with height (an isothermal layer) or actually increases (an inversion).

If the atmosphere is not so unstable, the shallow clouds will remain as fair-weather cumulus. These will probably grow a little in depth as the surface temperature increases during the day, then slowly decline and die as conditions cool toward evening.

Very often, the vertical extent of these clouds is limited by an inversion within the troposphere, rather than the tropopause itself. The culprit is the subsidence inversion that is found at about 1–1.5 km [about 1 mile] above the surface in anticyclones. The air sinks slowly in great depth to this level, while beneath it, the air is heated from the ground and churns over to produce cumulus cloud. The subsiding air above the inversion tends to be very dry and completely cloud-free.

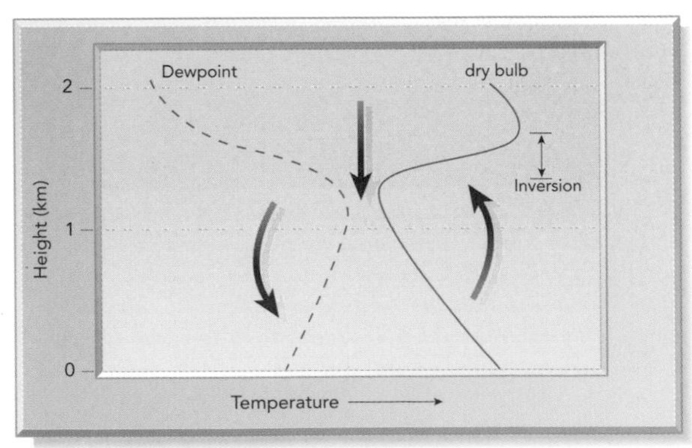

Low level overturning below a dry inversion, capped by subsidence

Very deep, vigorous cumulus: cumulonimbus with a shower

Scattered shallow cumulus

Very shallow cumulus, suppressed by subsidence above

High pollution concentrations occur underneath dry inversions

This means that although unstable bubbles of moist air ascend through the layer below the inversion, and often experience condensation, the tops of the cloud that forms become flattened at the base of the inversion. In these conditions, cumulus clouds adopt pancake-like forms and often spread sideways as continuous sheets of stratocumulus that cover most, or all, of the sky.

Rain spreading from the west...

Careful observation of how cloud forms evolve at a particular spot over some hours can also indicate the approach of a warm front.

The warmer, moister air rides up over colder, drier air to form an inclined warm front with a gentle slope of typically 1 in 100, or even shallower. The upgliding air within the warm sector is marked by condensation that occurs between low levels and the upper troposphere, several hundred kilometers ahead of the front's location on the surface.

Cumulus suppressed by subsidence

Middle-level cloud will occur 200–500 km [125–315 miles] ahead of the surface position, while high cloud will be seen between 500 and 1000 km [315–630 miles] ahead. Therefore, an observer watching the western sky would probably notice the approach of cirrus first of all. A careful watch on this type of cloud usually reveals that it is moving rapidly from the northwest (southwest in the southern hemisphere), around the upper ridge, rather than from the west whence the frontal system itself will likely approach.

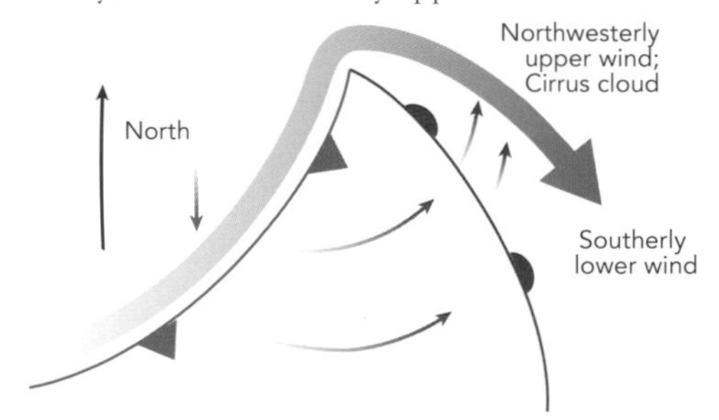

Cirrus approaching from the north-west in the high-level flow ahead of the surface warm front (south-west in southern hemisphere)

Cirrus (and contrails) can herald the approach of a distant warm front

Optical effects

If the cirrus is followed over an hour or so by cirrocumulus and/or cirrostratus, this usually indicates the start of a sequence of warm frontal cloud. Cirrostratus has a distinct appearance due to the way in which light from the Sun, or light reflected by the moon, is refracted by the ice crystals from which it is formed. A range of optical phenomena, including "Sun Dogs" (or Mock Suns) and rings around the

Cirrocumulus

Cirrostratus

Mock Sun in cirrostratus

Sun or moon, bear witness to the presence of cirrostratus. The former are seen best at times of low Sun, and appear as two localized bright patches either side of the Sun.

As time passes and the front moves closer to the observer, the cloud will lower, changing to altostratus and/or altocumulus cloud at middle levels. The rate at which these changes of cloud type occur will depend on how fast the depression is moving. Often, this will be about 15 m/sec, [50 ft/sec] although it can be as much as 25–30 m/sec [80–100 ft/sec] in vigorous winter storms. If the first cirrus is 1000 km [630 miles] ahead of a surface front traveling at 15 m/sec [50 ft/sec], it will presage the front's arrival by 16 or 17 hours.

Although there may be precipitation from the altostratus, it will not reach the surface. Altostratus cloud may make the Sun look opaque – it appears "as if through ground glass".

The altiform cloud will be followed by thickening, lowering cloud such as nimbostratus from which the first major precipitation may fall. Therefore, the first rain or snow reaches the ground a few hundred kilometers ahead of the surface front and typically lasts for a few to several hours.

The lowest cloud in the sequence occurs close to the surface front, and after this has passed, the sky tends to be laden with stratiform cloud. This can be very extensive or quite broken, especially over, and in the lee of, hilly areas.

In contrast to the general south-eastward movement of cloud at the highest levels ahead of the front, the surface flow will usually be *from* the south-east. The wind will often veer to south-westerly as the front passes.

Various types of cirriform cloud will often be seen in the sky, but they do not necessarily signify the approach of a front. The key to a successful D-I-Y forecast is being able to recognize the *sequence* of cloud formation. Other important indicators are wind direction, temperature, and pressure.

Altocumulus

Altostratus with scud below

Pressure changes

The approach of a frontal depression often leads to a fall in atmospheric pressure. However, this can be complicated by the pressure falling, rising or remaining the same in different regions of the system as it moves across the Earth's surface.

Generally, though, the progress of a warm front toward a spot is linked to a falling barometer. As the front passes, there is usually a weakening or leveling off of the pressure fall, followed by a rise in pressure as the cold front sweeps through. A fall in pressure usually foretells the imminent arrival of a depression or trough, while a rise is the expression of the gradual improvement of conditions associated with a high or ridge.

Warm sector

The region between warm and cold fronts is often marked by extensive low-layer cloud – especially on windward coasts – and occasional precipitation that is usually heavy shortly before the arrival of the cold front. Conditions are mild and damp in winter, and cool and damp in summer.

Sometimes, the extensive stratiform cloud breaks up inland, especially if over hilly areas. This means that sunny spells can develop, producing much higher temperatures.

Cold front

Harbingers of the arrival of a cold front are relatively few. The cloud formations, for example, will be behind the surface front because it tilts backward, away from the warm sector, trailing its arrival at the surface. Heavy downpours may herald the cold front, as may a veer in the wind direction, although this change is often most noticeable during and immediately after its passage.

Thus, the approach of a warm front is the easiest to identify by watching the sky.

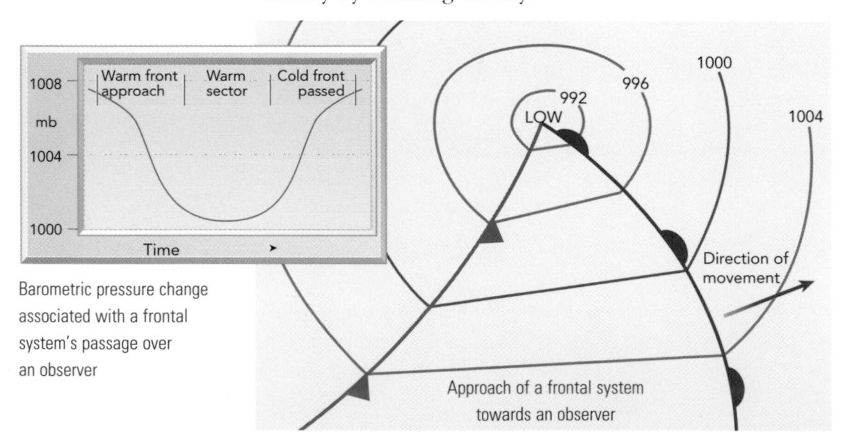

Barometric pressure change associated with a frontal system's passage over an observer

Approach of a frontal system towards an observer

chapter seven

Hazardous Weather

In many parts of the world, the most common natural hazard to affect society is the weather. Its impact may be very short-lived – from the disastrous transit of a tornado, to the potentially devastating passage of widespread gales that may last a day or so – or much more extensive, such as widespread flooding that may persist for weeks, or drought that may last for a season or longer.

The impact of hazardous weather can depend upon the economic health of the region or country affected. Inevitably, developing areas with poor infrastructure are hit far harder by events like hurricanes and drought.

However, in developed countries such extreme events are attracting more attention than ever before from insurance companies concerned with assessing the various risks. The predicted changes associated with global warming include the possibility that intense frontal storms in middle latitudes will become more frequent, while the inexorable increase in global sea levels will lead to an increased flood risk.

It is not only the incidence of natural disasters that may be on the increase, however, for there is also a global trend for population and wealth to be concentrated in cities that are frequently in high-risk areas – by the coast, for example. In addition, modern, massive urban complexes are probably more susceptible to disruption of their infrastructure than they were in the past.

Hurricanes

The term "hurricane" comes from the Spanish *huracan* and Portuguese *huracao*, which themselves are believed to originate from the Carib word *uracàn*, meaning "big wind". Similarly, typhoon is believed to originate from a Chinese dialect term *tai feng*, again meaning "big wind". "Hurricane", "typhoon" and "cyclone" are some of the names used regionally to describe the same feature – a tropical revolving

Cut-out view of the central features of a hurricane

storm that is typically 500–800 km [300–500 miles] across, which has a mean surface wind speed of 64 knots. If the wind speed is between 34 and 63 knots, the system is classified as a tropical storm, being given a name or number depending on the ocean basin over which it originated.

Where are they found?

These extremely hazardous weather systems occur most commonly across the low-latitude north-west Pacific and its '"downstream" land areas, where just over a third of the global total of such storms develop. The north-east Pacific averages 17%, while the North Atlantic typically sees 12% of the world total.

The busiest time for tropical cyclones in the northern hemisphere is between July and October, with a peak during August and September, partly because the sea surface temperatures are at their highest then. This feeds more water vapor into the systems through evaporation. Similarly, in the southern hemisphere, the peak season occurs when the sea is warmest, in January and February.

Tropical cyclone origin regions during a 20-year period

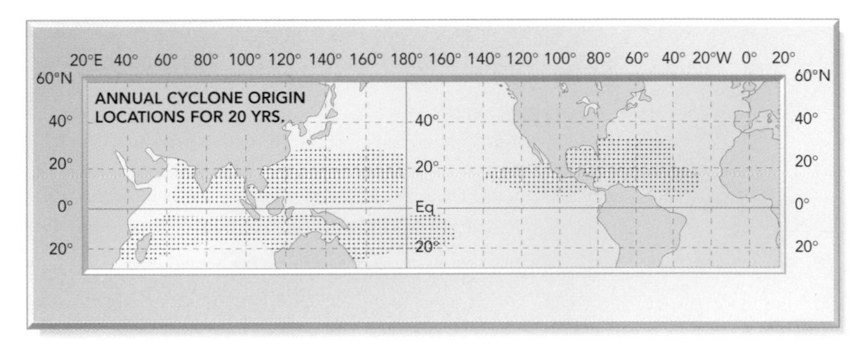

Where does the worst weather occur?

The damage produced by hurricanes and their regional cousins is not purely wind-related. Very high winds do occur, of course, where the horizontal pressure gradient is steep in the extreme, around the edge of the eye. Within the eye itself there is hardly any change of pressure across the surface. The eye is typically 20–30 km [15–20 miles] across and experiences deeply subsiding air with generally cloud-free skies. Surrounding it, however, is the eyewall cloud, which is like an upright cylinder and composed of extremely deep and vigorous cumulonimbus. It is across this zone that the worst winds and torrential rain occur.

Extremely strong winds and heavy rain will be encountered elsewhere within the circulation of a hurricane, especially in the spiral rainbands that are also composed of very deep cumulonimbus.

National Hurricane Center, USA
http://www.nhc.noaa.gov

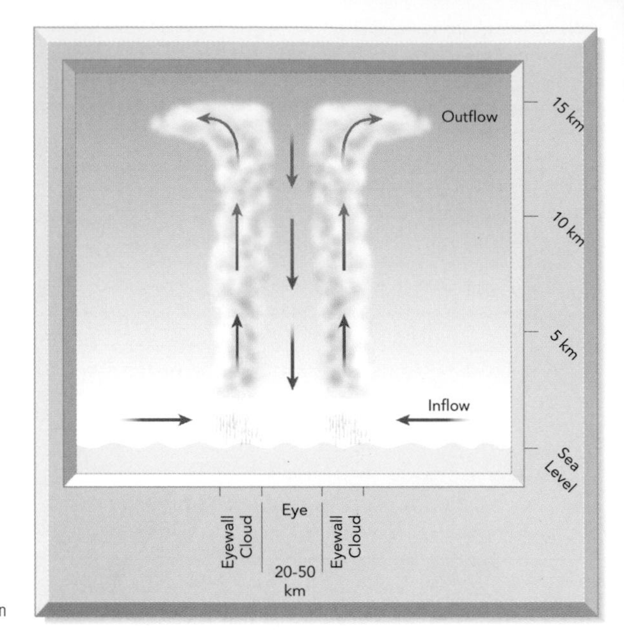

Vertical slice through central features of a hurricane/typhoon

Outflow aloft

At the top of a hurricane, the air spirals out in direct contrast to the inward swirling air in the lowest few kilometers of the troposphere. The strength and depth of this outflow plays a crucial role in determining whether the system will become more vigorous or weaken. If the mass of air being thrown out in the highest reaches of a hurricane is greater than the rate at which it is being supplied in the lowest kilometer above the sea's surface, the surface pressure will fall and the winds will probably increase.

Defining the strength

Although the traditional definition of a hurricane is Beaufort Force 12 winds ("air filled with foam, sea completely white

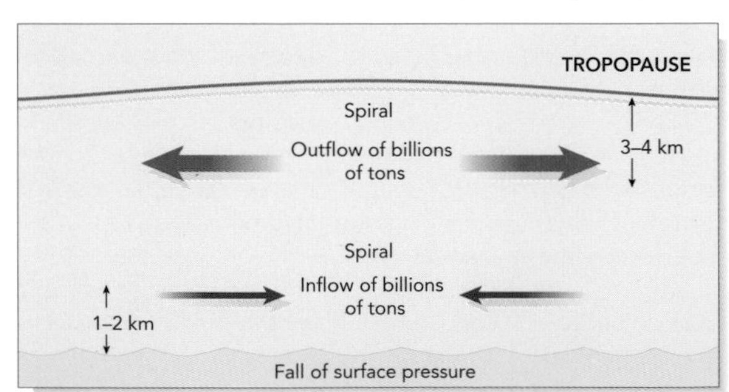

Link between surface pressure and upper/lower winds

with driving spray, visibility greatly reduced"), nowadays the Saffir-Simpson scale is used, especially along the East and Gulf Coasts of the United States. This scale ranges from 1 to 5 and refers to the magnitude of the average wind speed, and the storm surge plus the nature of likely damage. It is used as a quick means of informing not only meteorologists, but also the public, of the relative intensity of an approaching storm.

Saffir-Simpson Scale

Category 1 (weak hurricane)
65–82 knot winds; surge 1.2–1.5 m above normal water levels.

Category 2 (moderate hurricane)
83–95 knot winds; surge 1.8–2.4 m above normal water levels.

Category 3 (strong hurricane)
96–113 knot winds; surge 2.7–3.7 m above normal water levels.

Category 4 (very strong hurricane)
114–135 knot winds; surge 3.9–5.5 m above normal water levels.

Category 5 (devastating hurricane)
Over 135 knot winds; surge more than 5.5 m above normal water levels.

Between 1899 and 1980, the number of hurricanes experienced by the continental USA in each category was as follows:

Categories 1 and 2 – 82 hurricanes
Categories 3 and 4 – 54 hurricanes
Category 5 – 2 hurricanes

The two most devastating hurricanes to hit the USA during this period occurred on Labor Day, 2 September 1935, at Matecumbe Key, Florida, and on the coasts of Louisiana and Mississippi on 17 August 1969 (Hurricane Camille).

What's the damage?

The damage associated with the rare Category 5 hurricanes is tremendously costly, in both economic and social terms. It is defined as follows:

Most trees and signs blown down. Very severe and extensive roof, window and door damage. Complete failure of roof structures on most homes and many industrial buildings. Some large buildings suffer complete structural failure, while some smaller ones are overturned and may be blown away. Complete destruction of mobile homes.

Surge creates major damage to lower floors of all structures less than 5 m [16 ft] above mean-sea-level and within 450 m [1476 ft] of the shore. Low-lying escape routes are cut by rising water three to five hours before the storm center arrives. Evacuation of residential areas on low ground within 8–16 km [5–10 miles] of the shore may be required.

In the United States, deaths due to hurricanes have declined in recent decades because of improved forecasting and better levels of preparation for disaster. However, increasing coastal development from Texas to Maine is ensuring that there is a constant rise in the number of people who are vulnerable to the winds, torrential rain, and coastal inundation from the surge that accompanies a hurricane. Florida's population, for example, has more than doubled to 14.6 million since 1970. Although fatalities have tumbled this century, the cost of the damage has increased significantly.

Watches and warnings

The US National Weather Service issues "watches" and "warnings" as a matter of routine, to alert the public to the risk of an impending serious weather hazard. A "hurricane watch" means that a specific region faces the threat of hurricane conditions within 24–36 hours. It does not mean that evacuation will be necessary, but it implies that the population should be ready for this if a "warning" is issued. The latter occurs when severe weather has been reported or is imminent, at which stage everyone should take the necessary precautions.

Abandoned names

If a hurricane causes great damage, its name is never used again – so as not to tempt fate, perhaps. This is the case for David and Frederick (1979), Allen (1980), Alicia (1983), Elena and Gloria (1985), Gilbert and Joan (1988), Hugo (1989), Bob (1991), Andrew (1992), and Mitch (1998).

The costliest

The top ten costliest hurricanes for the United States between 1900 and 1996 are:

		category	damage cost ($m)
Andrew	1992	4	26,500
Hugo	1989	4	7,000
Fran	1996	3	3,200
Opal	1995	3	3,000
Frederic	1979	3	2,300
Agnes	1972	1	2,100
Alicia	1983	3	2,000
Bob	1991	2	1,500
Juan	1985	1	1,500
Camille	1969	5	1,420

Andrew was the most expensive natural disaster ever to strike the United States. It crossed the Bahamas as a category 4 hurricane before moving on to southern Florida. It produced a trail of damage over five days from 23 to 27 August 1992.

Its eye moved onshore in Dade County – the central surface pressure of 922 mbar was the third lowest this century at landfall. The maximum sustained wind (averaged over 1 minute at a height of 10 m [33 ft]) was 220 km/h [139 miles/h] gusting to 265 km/h [167 miles/h] at landfall. It caused 15 deaths in the county and left about 250,000 people homeless.

Andrew took four hours to cross the Florida peninsula during which time it weakened to a Category 1 hurricane. However, once free to move across the warm waters of the Gulf of Mexico, it regained most of its vigor and had a second, less devastating, landfall near Morgan City on the coast of Louisiana. It then moved north inland and was downgraded to a tropical storm within 10 hours.

1998 – a busy North Atlantic season

The summer of 1998 was the first since 1892 that the North Atlantic experienced four hurricanes simultaneously. They were Georges, Ivan, Jeanne, and Karl. Luckily, the last three roamed across the ocean, although they did pose a threat to shipping. Ex–Karl ended up producing some wet and windy weather across parts of western Europe during late September. It is not uncommon for the remnants of hurricanes to rove that far, especially in late summer and early autumn.

Georges was another matter. On the backs of Bonnie, Danielle, Earl, and Frances, all of which produced damage at various points along the US Atlantic seaboard and inland, Georges was the worst for fatalities. It rampaged through the Caribbean, destroying virtually all the Dominican Republic's crops. It struck north toward the Mississippi delta region, provoking the evacuation of some 1.5 million inhabitants from New Orleans and adjacent areas. The city is around 2 m [7 ft] below sea level and is protected by huge levees and drainage channels built to withstand a Category 3 hurricane. Georges moved sluggishly toward the coast at about 9 km/h [6 miles/h], while its winds were roaring around its center at approximately 160 km/h [100 miles/h]. It produced some 200 mm of rain at Pensacola in Florida, while other areas were predicted to receive four times that amount.

Type	Category	Line Color
Depression	TD	
Tropical storm	TS	
Hurricane	1	
Hurricane	2	
Hurricane	3	
Hurricane	4	
Hurricane	5	

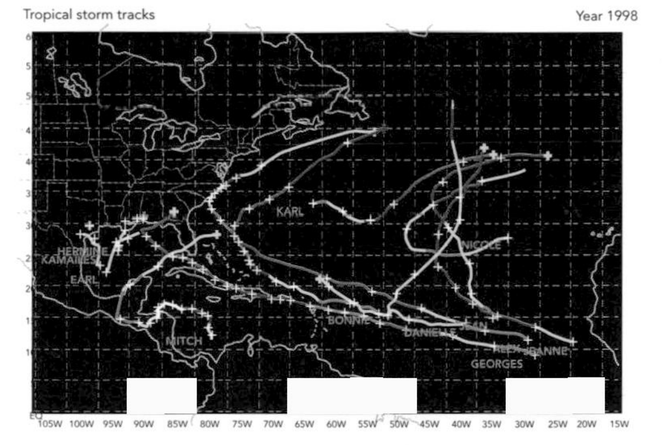

Tropical storm tracks over the North Atlantic in 1998

The eye made landfall near Biloxi in Mississippi with 165 km/h [104 miles/h] winds. 602 deaths were directly attributed to Georges across many Caribbean islands.

The 1998 season was characterized by an inactive period that lasted until mid-August, when the start of the season's stream of disturbances occurred. Bonnie was the first of significance, running into the Bahamas before slowing and becoming stationary off the south-eastern shores of the United States. It moved slowly into the Carolinas, then headed northeastwards to skirt the coast for many hundreds of kilometers.

Bonnie was followed by Charley, which attained the status of a tropical storm. The center of this disturbance crossed the Texas coast near Corpus Christi to bring rainfall to many areas of drought-ridden central and southern Texas. However, people were drowned in and around Del Rio on the Rio Grande, where some 500 mm [20 inches] of rain fell in the course of a day or two.

Hurricane Mitch

The 1998 season reached its tragic climax with the appearance of Hurricane Mitch in October. It was the strongest ever such system to be observed for the time of year: it formed on the 21 October about 580 km [365 miles] south of Jamaica. Early on 24 October, it had attained hurricane intensity and deepened rapidly to reach a minimum surface pressure estimated to be 905 mbar on 26 October. Its maximum sustained surface winds were thought to be near 285 km/h [180 miles] while offshore, but close to Honduras.

After moving around near the coast of Honduras for some time, it made landfall on the morning of 29 October then moved south across Honduras and Guatemala during the last two days of the month, producing prolonged and extremely intense downpours as it did so. The devastating floods all but completely ruined the economies of those countries and was the cause of around 11,000 fatalities. It is estimated that not since 1780 have so many people died in a hurricane disaster in the Caribbean/Gulf region.

Hugo's life story

One of the worst hurricanes of recent times was Hugo, which occurred in 1989. It swept across the Caribbean, hitting Guadeloupe on the night of 16–17 September. At one point, average surface winds reached 210 km/h [132 miles/h]. Homes were destroyed, and all electrical power was cut off. The next day, 18 September, Hugo tracked north-westwards toward Puerto Rico. To spur the inhabitants of that island into action and move them into hurricane shelters, the radio broadcast, shown on the right, was issued at 0900 local time.

National Climatic Data Center, USA
http://www.ncdc.noaa.gov/ol/reports/mitch/mitch.html

'If the eye of Hugo moves across Puerto Rico as forecast, we can expect a 50-mile-wide path of extensive and extreme damage to occur. The storm surge will decimate the coastal section where it comes onshore. Then, hurricane-force winds will destroy structures and uproot trees. Roofs could be removed and loose objects will become lethal airborne projectiles.'

Hugo east of Florida

The surge

Major killers in Puerto Rico are the mudslides that occur after heavy rains; luckily, no such fatalities were reported during Hugo. The surge associated with a hurricane is caused by the sea's surface becoming domed beneath a low-pressure system. In contrast, the surface is "squashed" down by high

Effect of barometric pressure on the sea surface

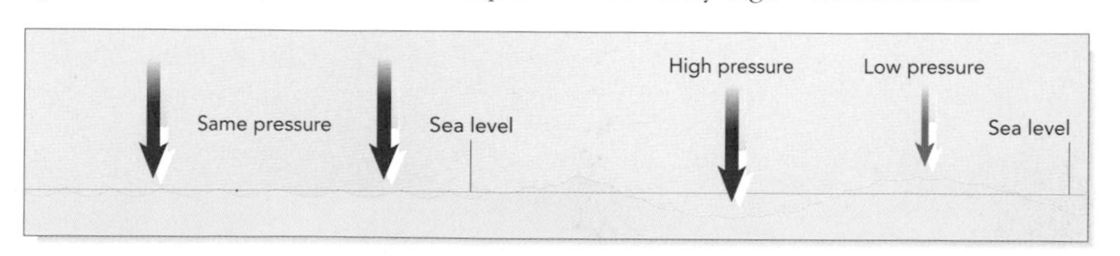

pressure. This response of the sea's surface is called the inverse barometer effect, and for a 1 mb change in air pressure, the sea level will rise or fall by roughly 1 cm [0.4 of an inch]. Thus, a very deep hurricane will be mirrored by a traveling region of elevated sea surface. To compound the impact of the surge, the hurricane's direction of motion adds to its height, as does the force of the wind on its forward right quadrant. The same applies on the forward left quadrant of such systems in the southern hemisphere.

Caribbean devastation

Rainfall totals on Puerto Rico during the passage of Hugo varied from 125 mm [5 inches] at Arecibo on the north coast to 344 mm [14 inches] at Lower Rio Blanco in the northeastern mountains. Staff at the San Juan Weather Service

Hurricane Hugo approaching Georgia and South Carolina

Forecast Office ran the risk of shocks from electrical equipment because heavy rain was driven through the building's hurricane shutters. Windows bowed under the winds and were kept in place by placing filing cabinets against them!

Hugo was responsible for the death of one person in Antigua and Barbuda, six in the US Virgin Islands, one in St Kitts and Nevis, 11 in Guadeloupe, ten in Montserrat, and 12 in Puerto Rico. After devastating the Caribbean, it increased speed from about 11 knots to 20 knots as it bore down on the Carolinas during 21 September.

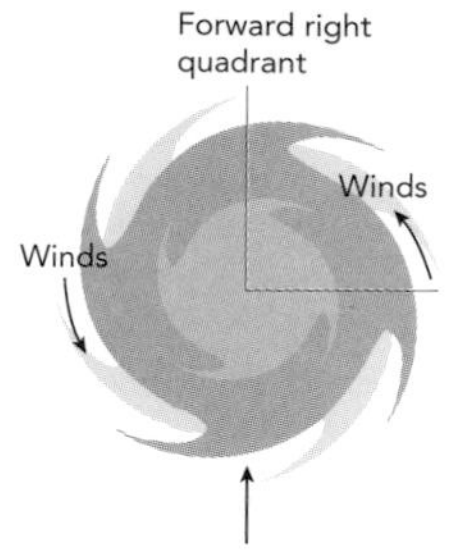

Region of worst impact: the forward right quadrant (left in the southern hemisphere)

Warning from Miami

The National Weather Service Office in Miami issued the following advisory at 10.00 pm on Thursday 21 September:

"... extremely dangerous hurricane Hugo bearing down on South Carolina Coast....

'A hurricane warning remains in effect from Fernandina Beach, Florida, northwards to the Oregon Inlet, North Carolina, including the Pamlico Sound. A tropical storm warning and a hurricane watch continues from the north of the Oregon Inlet to Cape Henlopen, Delaware, including the Albemarle Sound and Chesapeake Bay. Reconnaissance reports and satellite pictures indicate that Hugo is a Category 4 hurricane on the Saffir-Simpson Hurricane scale. On the present course ... the center is expected to cross the South Carolina coast near Charleston shortly after midnight tonight near high astronomical tide. All precautions should be completed as gusts to near hurricane force are now being reported on the South Carolina coast.

At 10 pm EDT 0200 UTC the center of Hugo was located near latitude 32.2 north longitude 79.3 west. This position is about 60 miles south-east of Charleston, South Carolina, or about 105 miles south of Myrtle Beach. Hugo is moving towards the north-west near 22 mph and this motion is expected to continue with a gradual turn towards the north in the next 12 hours. Maximum sustained winds are near 135 mph and little significant change in strength is likely prior to landfall. Hurricane force winds extend outward up to 140 miles from the center and tropical storm force winds extend outward up to 250 miles. The minimum central pressure reported by an Air Force reconnaissance plane is 938 mb (27.70 inches). Storm surge flooding of 12 to 17 feet above normal astronomical tide levels can be expected near and to the north of where the center of the hurricane crosses the coast. Details of tides and actions to be taken are included in local statements being issued by National Weather Service and Local Government Officials. Rainfall amounts of 5 to 10 inches can be expected in the path of the hurricane. A few tornadoes are likely in portions of South and North Carolina tonight."

Radar echoes (rainfall intensity) from Charleston, South Carolina around the passage of Hugo's eye

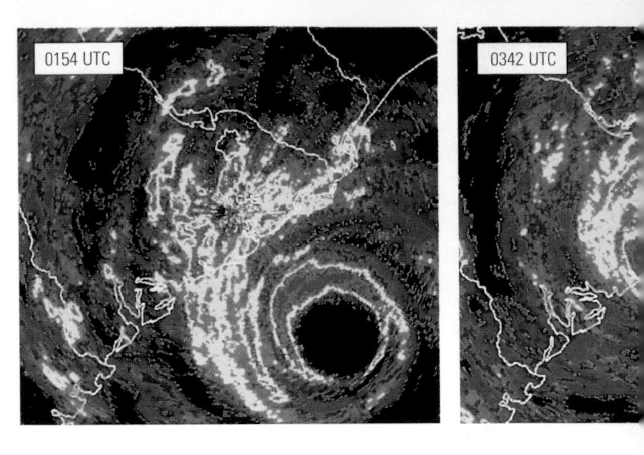

The observations confirmed the quality of the forecast. The eye and spiral rainbands showed up clearly on the Charleston precipitation radar, while the surge maximum and wind speed maximum both lay to the right of the track of the eye.

The fact is – and will remain so – that the United States is equipped to organize mass evacuation very effectively indeed. This is not the case in many Caribbean states, where the infrastructure and means of communication are often poorer.

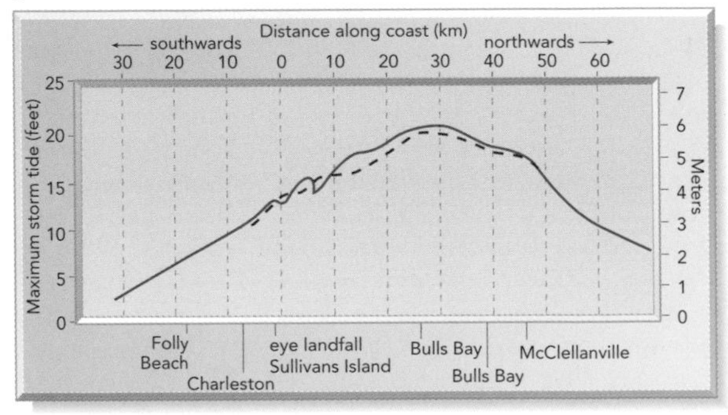

Predicted (solid line) and observed (dashed line) surge height for Hurricane Hugo along a stretch of South Carolina coast

Severe convection

Although tropical cyclones are quite large features, many of the terrible conditions they produce are related to extremely deep thunderstorms embedded in their spiral rainbands and eyewall cloud. In effect, the large-scale pattern is made up from a significant number of smaller-scale cloud features.

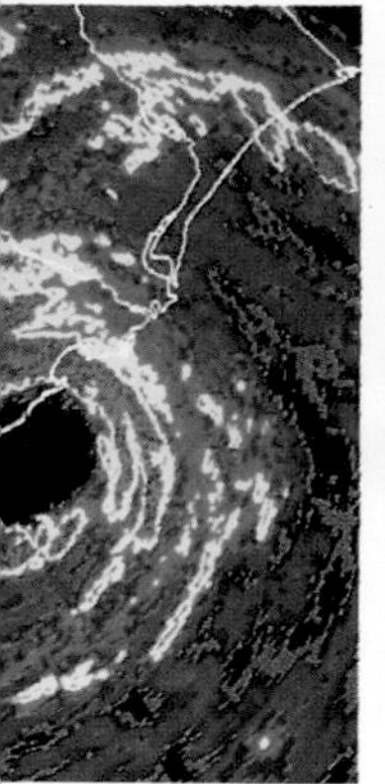

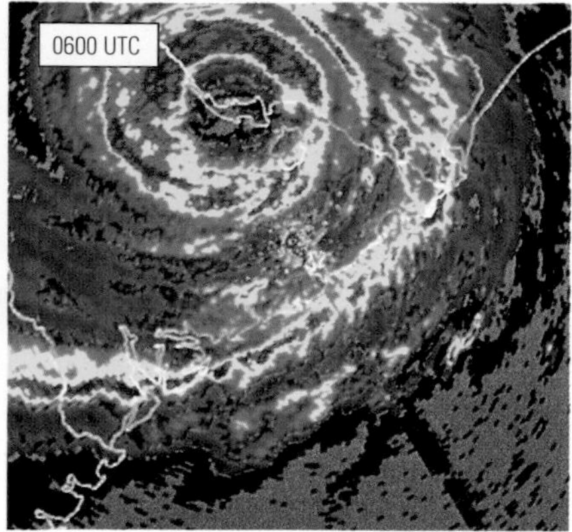

"Severe convection" refers to troposphere-deep cumulus cloud that produces dangerous weather like lightning, hail, extreme gustiness, and the occasional tornado! In the USA, a severe thunderstorm is defined as one that produces winds on the ground of at least 93 km/hr [59 miles/h] and hailstones that are at least 20 mm [0.8 of an inch] in diameter.

Hail

Hail comprises of large pieces of ice that form within, and fall from, a cumulonimbus cloud. Such deep convective clouds are characterized by strong updraughts and down-draughts.

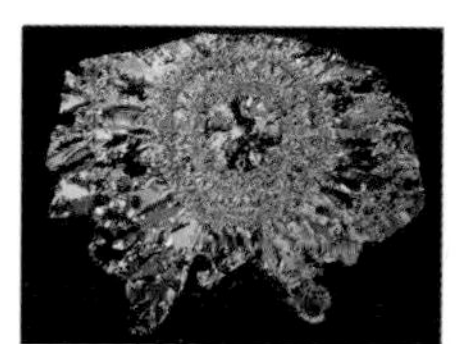

Coffeyville, Kansas: Hailstone in polarized light, illustrating a number of ice layers

The hailstones grow from graupel, which act as a nucleus, becoming larger due to the accumulation of supercooled water droplets as they are borne upward on the rapidly ascending air. It is not uncommon for golfball-size hail to occur in the United States during the summer. Such hailstones probably will have been up and down through the same cloud a few times – over the course of ten minutes or so – before they acquire sufficient layers of ice to be heavy enough to fall out of the cloud and on to the surface.

It is possible to count the number of ice layers in a large hailstone and thus gain an idea of the number of ascents it has made in the water-rich updraught.

Damage

It seems paradoxical that the largest hailstones are observed during late spring/early summer. Although surface heating is instense, this is also the season when overrunning cold air at upper levels can lead to deep overturning motions within the troposphere, expressed by deep convection; water vapor concentrations are also high due to enhanced evaporation. These conditions reach their height across the North American high plains, stretching from Texas to Alberta.

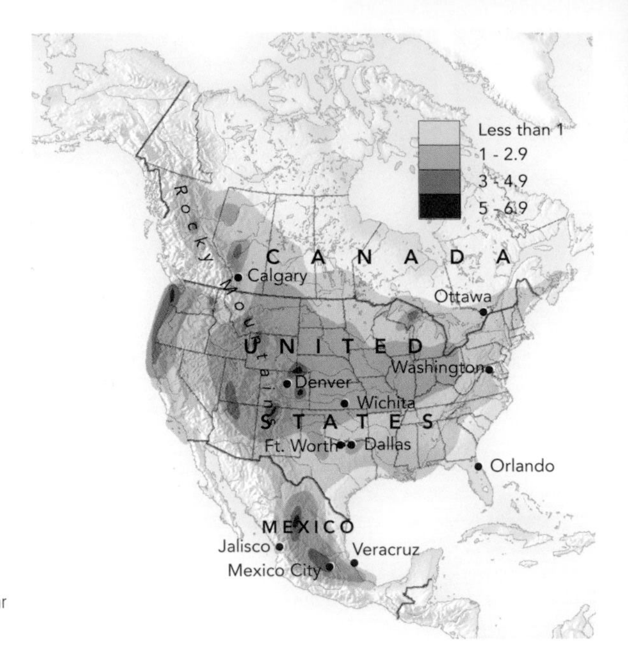

Average number of days per year with hail, across USA, 1955–95

The number of hailstorms reported in the USA peaks in May and June, although April and July are also busy months. Damage to property ensues when hail reaches some 20 mm [0.8 of an inch] in diameter – note that an average May and June will each experience over 2000 storms that produce hail with a diameter larger than 50 mm [2 inches]!

The largest authenticated hailstone to fall in the USA was found at Coffeyville, Kansas, in September 1970. It weighed 757 grams [1665 lbs] and had a diameter of some 14 cm [6 inches]! A cross-section of this particular hailstone revealed the distinct layering within it.

In the USA, hail causes about $100 million worth of damage every year, to both property and crops. The costliest hailstorm to date is one that occurred in Denver, Colorado, on 11 July 1990, when the damage totalled $625 million.

Downbursts

The intense precipitation that typifies mature thunderstorms' updraughts evaporates as it falls, which can create an evaporatively-chilled volume of air that descends from middle levels to the surface as a downdraught. When it reaches the surface, it flows out sideways as a "density current" of cool, blustery air. If particularly intense and localized, it is known as a downburst or microburst depending on its width.

The sudden change of wind direction and speed as they spread out across the surface is dangerous to aircraft that are landing or taking-off, since aircraft must land and take-off into the wind and all "flaps" must be set properly. In the USA where severe convective clouds are particularly high risk in the spring and summer, all major airports are equipped with Doppler radar to sense the location and approach of the threatening gustfronts. Short-term forecasts are issued to warn of the passage of these features across airports so that landing and taking-off operations can temporarily cease.

Thunder and lightning

These two always go hand-in-hand. Lightning is a massive electrical discharge between one cloud and another, from a cloud into the air, or between a cloud and the ground. The way in which electric charge becomes separated within thunderclouds is not completely agreed upon – the main theory being that when hail and graupel fall through a layer of supercooled water and ice crystals that form the cloud, there is a transfer of positive charge from the slightly warmer hail to the colder cloud particles. The larger hail becomes negatively charged, accumulating such charge in the lower layers of the cloud. Conversely, the water and ice crystals gain positive charge and tend to accumulate in the upper reaches of the cloud on the updraught.

Discharge

As the lower negative charge grows with the evolution of the cloud, it induces a region of positive charge below it on the surface, which moves along beneath the drifting cloud. This positive charge tends to be concentrated on objects that

Evolution of a lightning stroke from stepped leader to return stroke

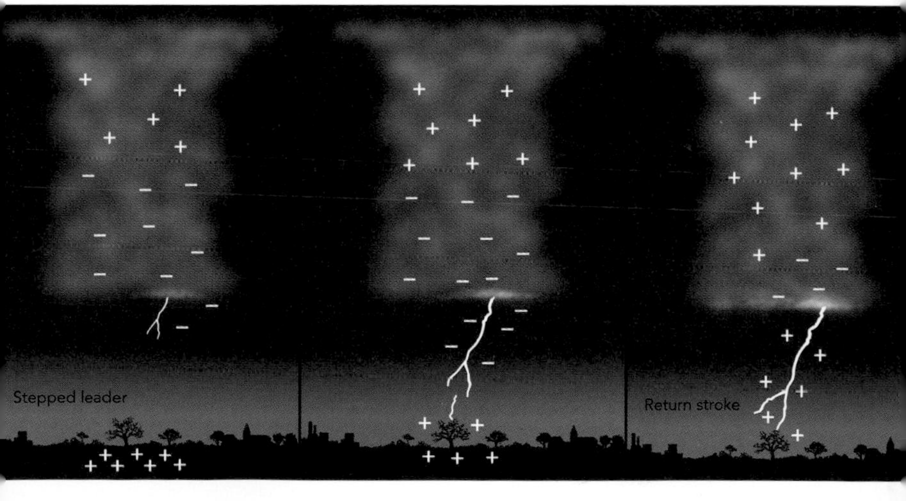

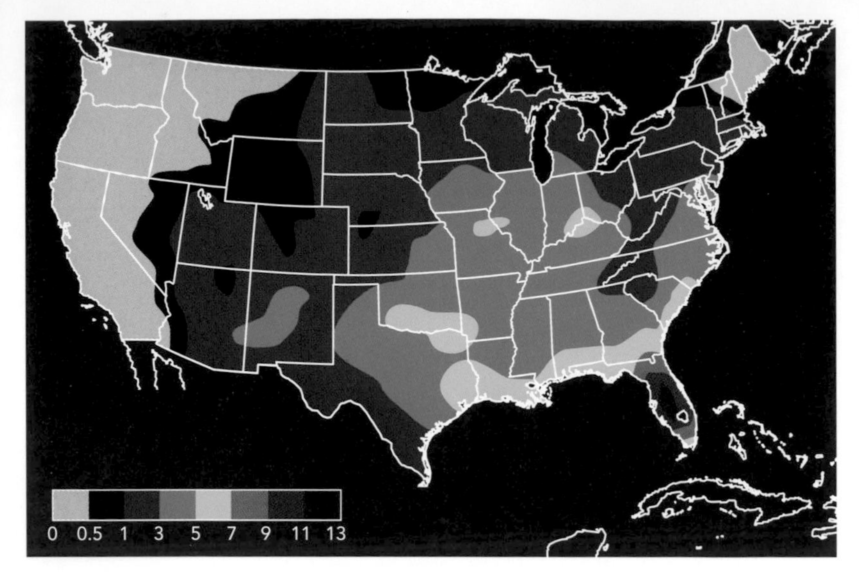

Cloud to ground flash density in 1991 (flashes per square km)

protrude from the surface and which are relatively isolated. Although dry air is quite a good electrical insulator, the potential difference that will grow under the right conditions is so enormous that a massive discharge is unavoidable. A difference of about 1,000,000 volts/m is typical and will lead to a current of up to 100,000 amperes.

Flash

Lightning that reaches the ground develops within the cloud, where electrons move rapidly down toward the base of the cloud, but in a stepped fashion. Every discharge runs for 100 m [330 ft] or so, then halts for about 50 millionths of a second before continuing downward. This process is continued as an invisible stepped "leader" until, near the ground, the potential gradient is so large that an upward positive current leaves the surface from tall objects such as trees and buildings. Once these two currents meet, electrons flow down to establish a channel that is used by a larger return stroke. This massive, brilliant upcurrent is what we see, and it lasts typically for one ten-thousandths of a second.

Fires

In western USA, many fires, especially in forests, are started by lightning. During a recent ten-year period, over 15,000 such fires occurred across the whole USA. These resulted in damage worth several hundred million dollars and the destruction of some two million acres of forest. In addition, on average, lightning causes 93 deaths and 300 injuries a year in the USA.

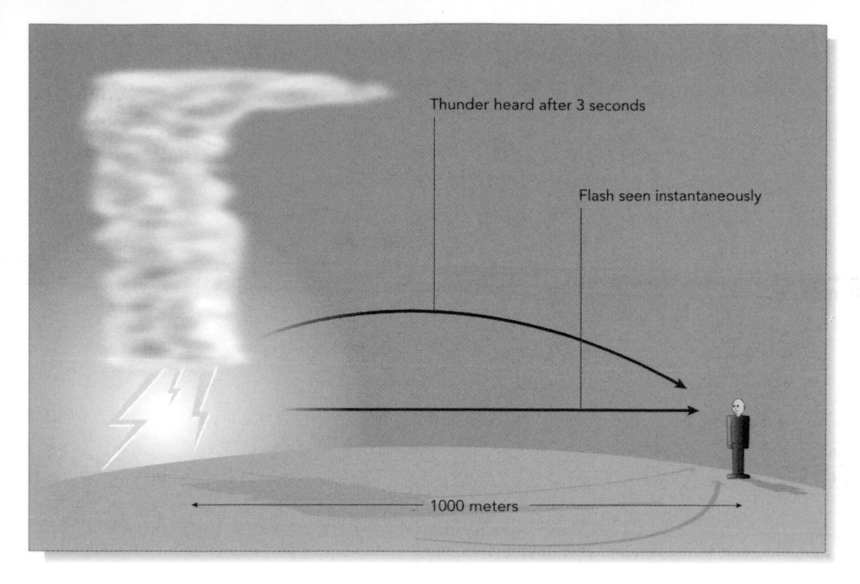

Delay between seeing lightning and hearing thunder

The incidence of cloud-to-ground lightning flashes across the USA in 1991 displays a marked maximum in the humid, hot south-eastern region of the country. Of course, there are flashes over the sea, but the vast majority are associated with deep convection over land. Florida had a notable maximum in 1991 with between 11 and 13 per square kilometer throughout the year, most of which would have occurred in the hotter months. In general, the higher incidences of lightning flash are confined to the region east of the Rockies, where warm, humid air from the Gulf is an important ingredient in the formation of thunderstorms.

Flash bang

Only about one in five lightning strokes are from cloud to ground. Each instantly heats the channel of air through which it flows by about 30,000°C [55,000°F]! This means that the air expands incredibly quickly and very dramatically to produce a shock wave, which travels away from the lightning stroke at the speed of sound.

The light from lightning reaches our eyes instantaneously, but the sound of thunder emanates from it at about 330 m/sec [1100 ft/sec]. This forms the basis for a rule that we can use to estimate our distance from the ground stroke.

Every second between the flash and the thunder indicates a distance of about 330 m [1100 ft]. Therefore, a pause of three seconds means that the lightning hit about 1 km [0.6 miles] away. This rule holds good for distances of up to 5 km [3 miles]; beyond that, we do not often hear thunder, because the sound is absorbed and refracted by the air.

Tornado and base of parent cumulonimbus cloud

Tornadoes

Tornadoes are always linked to a parent cumulonimbus cloud. Although many people confuse them with hurricanes, they are actually very much smaller. The most frequently observed size of a tornado's damage path is about 50 m [160 ft] wide with a track of 2–4 km [1–3 miles]. However, the largest damage swathe can exceed 2 km [1 mile] in width, and the narrowest, 10 m [30 ft]. Tornadoes are most notorious in North America, but, with the exception of Antarctica, can occur in other continents.

A single tornado can last from a few seconds to over an hour. The typical duration is around five minutes. To be defined as a tornado, the vortex of rapidly spinning air must be in contact with the ground. This means that if debris is visible, even if there is no obvious funnel cloud, a tornado is present. The surface wind speeds are estimated from the nature of the damage produced – it is believed that they can reach up to 460 km/h [300 miles/h] in extreme cases.

The F-scale

The Fujita (or F-) scale is used to convey tornado intensity to the meteorologist. This was devised by the meteorologist Theodore Fujita in the late 1960s and is set out below:

F0	up to 115 km/h	light damage
F1	116–179 km/h	moderate damage
F2	180–251 km/h	considerable damage
F3	252–330 km/h	severe damage
F4	331–416 km/h	devastating damage
F5	over 417 km/h	incredible damage

National Severe Storms Laboratory
http://nssl.noaa.gov

The cost

The worst American tornado for fatalities was the Tri-state outbreak of 18 March 1925, when 689 people across Missouri, Illinois, and Indiana lost their lives. It is a fact, however, that of the top 25 US killer tornados, none have occurred since 1953, as a result of the improvement in forecasting the severe convection with which they are associated, and the method of conveying watches and warnings to the public.

The value of damage does highlight the power of recent tornadoes, however. Three very costly ones were those of 6 May 1975 ($400 million in Omaha, Nebraska), 10 April 1979 ($400 million in Wichita Falls, Texas), and 3 October 1979 ($200 million in Windsor Locks, Connecticut).

The worst tornado outbreak ever to hit Oklahoma occurred on 3 May 1999. Tornadoes are not unknown in the state, but this event was one of extreme severity and one that unfortunately affected heavily populated areas. More than 50 twisters ran across central Oklahoma that day; 40 people perished in, and to the south-west of, Oklahoma City. There were also outbreaks that day in parts of north Texas, eastern Oklahoma and south central Kansas. Five people died in Wichita.

One tornado in particular became the major killer, spawned by a massive supercell storm that had already produced a crop up to F3 intensity in the countryside some distance to the southwest of Oklahoma City. It touched down close to the small town of Chickasha as an F1, then sped northeast menacingly toward the metropolitan region. On its way it geared up to F4 intensity, declined to F3, then reintensified to a rare F5 over the community of Bridge Creek where 680 homes were totally destroyed. A second, more destructive, 1.5 km [1 mile] wide tornado grew near to the Canadian River during the evening. It quickly attained F5 status and devastated Moore, a suburb of Oklahoma City. The destruction here was unbelievable; 1225 houses and 274 apartments were razed along with 50 businesses, two schools, and churches. In addition, 4000 to 4500 homes were damaged. The estimated total cost borders on $1000 million.

Timely warnings were issued for all the affected areas. The problem with the most severe tornadoes was their unusual intensity. Even solidly built houses couldn't withstand their ferocity.

Predicting tornadoes

It will probably never be possible to predict the precise location and timing of a tornado, even one day ahead. Currently, forecasters are well aware of the risk of severe convection one day ahead, and based on this knowledge, they issue advice for areas that may encompass a few US states, or regions within states, indicating the increased risk to such places.

The situation is monitored by routine surface observing stations, although they do not provide the fine detail of precipitation. Precipitation radars map the extent and intensity of rainfall – and indicate the location and movement of severe storms. However, such storms only register on this type of radar once they have started up. The other type is Doppler radar, which maps the regions of convergence and divergence in the lower atmosphere. Regions of convergence are the risk areas and are often useful precursors of deep convection.

Once severe storms have developed, their changes in motion and intensity can be monitored virtually continuously from the ground. Rapid scans (up to an image a minute) from the US geosynchronous weather satellites are permitted when conditions indicate a high risk of severe convection.

1995 – a bad year for the United States

The range of weather-related dangers is highlighted by analysing the record from 1995 across the USA. The average annual number of deaths caused by weather events there between 1986 and 1995 was 485. In 1995, the total was 1364; the list below is not exhaustive.

Event	**Deaths**	**Damage** (millions of $)
Lightning	85	33.1
Tornado	30	410.8
Thunderstorm winds	38	745.1
Hail	2	1449.3
Cold	22	633.4
Heat	1021	456.9
Flash flood	60	902.4
River flood	20	348.1
Hurricanes/tropical storms	17	5932.3
Snow/blizzard	11	108.5
High winds	46	121.0

A detailed examination of these figures reveals that there were 948 deaths in July followed by 171 in August.

A good deal of these summer fatalities occurred in Illinois, where 629 died during a baking heatwave between 11 and 27 July. Many died in Chicago, and in Milwaukee, Wisconsin, where the state total was 89. Pennsylvania saw 104 deaths.

Of these deaths, 89% occurred in solidly-built homes, and 67% of those who succumbed were in the 60–89 year age range. This group is very vulnerable, as are the very young. Many older people died in Texas and Oklahoma during the persistent extreme heat of summer in 1998.

Mid-latitude frontal storms

Significant strides have been made in improving our understanding of, and ability to predict, severe frontal systems that affect western Europe (or geographically similar areas) since the famous event of 16 October 1987. Contrary to popular belief, it was not a hurricane.

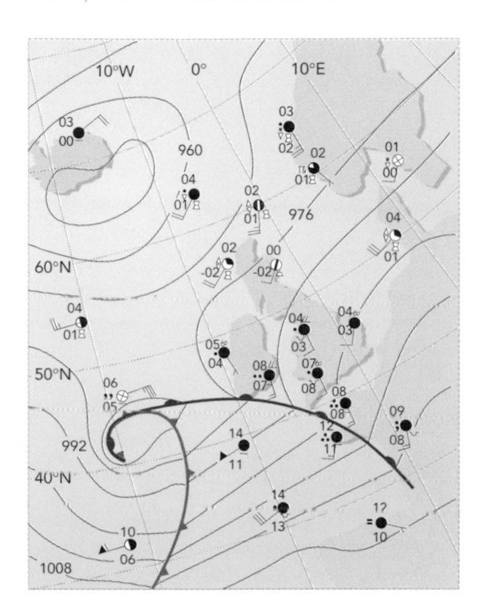

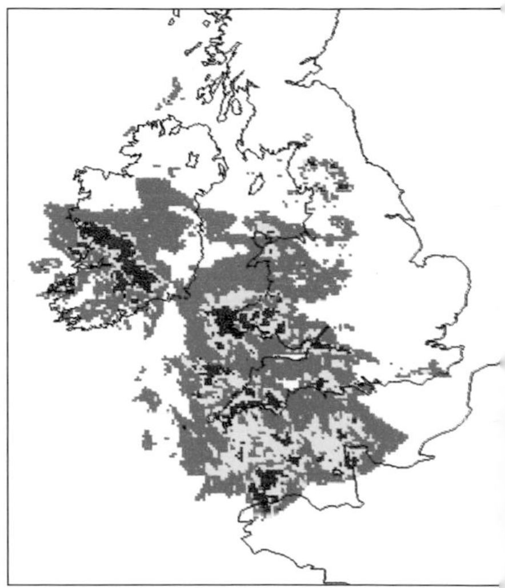

Surface weather map and radar rainfall map, 00 UTC, 25 January 1990

Frontal depressions: The Burns' Day Storm

Occasional severe weather warnings are issued in the UK every winter, often in relation to the passage of a deep frontal depression (very low central pressure) that can produce widespread wind damage and flooding. Sometimes, however, there may be an unusually extreme development of a rapidly deepening depression with winds of dangerous strength. Since the October storm of 1987, the most damaging frontal system to hit Britain and Ireland has been the so-called Burns' Day Storm that rushed through on 25 January 1990.

Extremely wet and warm

The British winter of December 1989 to February 1990 was the wettest and warmest since 1914–15. Across Scotland, for example, the three months from January to March 1990 each recorded a precipitation total that fell within the ten highest

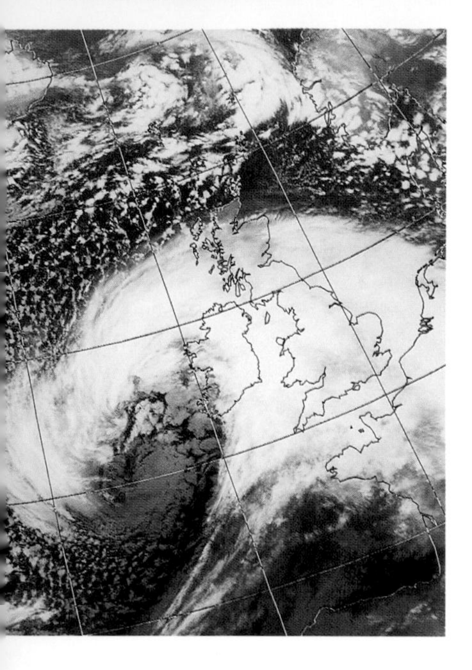

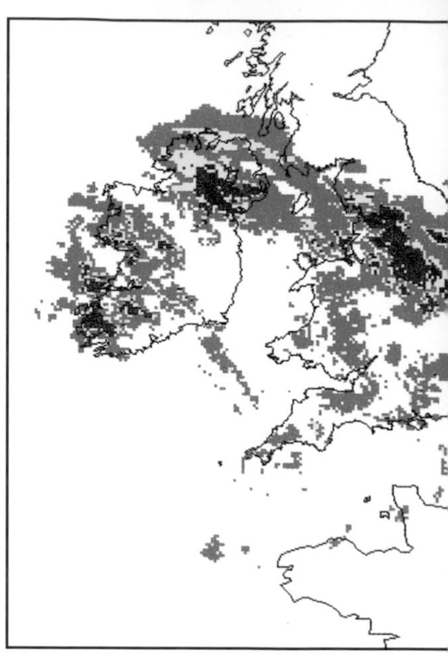

Satellite thermal infrared image (0320 UTC) and radar rainfall map (0400 UTC), 25 January 1990

monthly values in Scotland since 1869. This period produced an average rainfall for Scotland of 791 mm [31 inches] – the highest on record.

Groundwater levels for the UK in February 1990 had no modern precedent; run-off rates were enormous, and before the end of the first week of that month, most rivers were in spate. At Chilgrove Observation Borehole in West Sussex, the level rose by 40 m [130 ft] in eight weeks from a near-record low level in early December.

In the western headwaters of the Tay, Britain's largest river in terms of discharge, 518 mm [20 inches] were recorded in the 25 days leading up to 4 February. Moderate rainfall after the this date, combined with extensive snowmelt in the mild weather, produced a peak flow at the Tay's Ballathie Gaging Station of 1750 m^3/sec – the highest recorded UK rate.

The Burns' Day storm was only one component of this dramatic winter, but was well predicted before its arrival.

The storm approaches

The depression and its attendant fronts were just approaching Ireland, Britain, and France at midnight on 24 January, the warm front stretching from south-west Ireland, through the extreme west of Cornwall, to a point near Bordeaux, France. The radar network mapped the extent of the rain preceding the warm front as far ahead of the surface front as the southern Pennines and north Midlands.

It sweeps across

During the night, the system ran across the British Isles, and at 0329 UTC on 25 January, the forward edge of the warm front's cirrus was well across the North Sea, while the cold front stretched away south-westwards from the Celtic Sea. By 0400 UTC, the rain had also moved on rapidly, although the sharp edges over the North Sea are probably expressions of the radar's limit of view rather than a real boundary. By 1200 UTC, the low's center lay over northern England, and the frontal cloud mass had left most of the British Isles at the time of the satellite image at 1324 UTC. Much wind damage occurred across Britain.

In the UK, we never see hurricanes. Winds may reach hurricane force on the Beaufort Scale, but they are produced by frontal depressions, not systems that have an eye and spiral rainbands that produce torrential rainfall.

What does happen from time to time, mainly in the late summer and fall, is that a system that began as a hurricane brings strong winds and heavy rain to western Europe. By the time it reaches these shores, however, its tropical characteristics will have died.

Forest and bush fires

Many regions of the world are susceptible to forest and bush fires, particularly those that experience a significant dry season during the year when conditions are hot. Some weather services advise forestry departments regarding the type of weather that increases the risk of such fires.

The occurrence of an El Niño (see Chapter 9) puts tropical countries in the western Pacific region at a significantly higher risk than normal because of the enhanced, and prolonged, subsidence experienced there. The vast pollution event in Malaysia and Indonesia during September 1997, devastating bush fires in Australia in late 1997 and early 1998, and widespread forest fires in Borneo during April 1998 were all partly related to the 1997–98 El Niño.

Aussie bush fire

On 15 December 1997, Australian firefighters fought to control a massive bush fire in the north-west of the state of Victoria that had destroyed almost 4000 hectares of natural scrub in the Murray-Sunset National Park. Some 45 fires had started on the previous day, due largely to lightning strikes. One fire co-ordinator stated that, "With the number of fires for the current fire season exceeding those for the same period during the 1982–83 season, the indication is that we are in for a grim summer." Interestingly, 1982–83 was the last time the world saw a strong El Niño.

During the period 3–9 January 1994, there were five days during which the "McArthur Forest Fire Danger index' exceeded the value of 50, indicating extreme fire danger. The combination of circumstances producing such "ripe" conditions were a prolonged drought (a deficit of over 100 mm [4 inches]), a temperature of 35°C [95°F], a relative humidity of 15%, and a mean wind speed of 33 km/h [21 miles/h].

In January 1994, almost all of coastal New South Wales experienced weather that produced a high fire risk. Between 27 December 1993 and 16 January 1994, over 800 fires destroyed about 800,000 hectares of vegetation, and over 200 houses – mainly in suburban areas, including Sydney's. Two firefighters and two members of the public died in the fires during this period. This loss was not as dramatic, however, as two other major events. On Ash Wednesday 1983, 76 died and over 2500 properties were burned down in the Adelaide Hills and outer Melbourne. Around Hobart, Tasmania, 62 died and over 1300 buildings were destroyed in 1967.

An important message for town planners and local authorities is that these disasters will occur at regular intervals if urban development is permitted to encroach on surrounding forests and bushland. Despite this, such development of this kind continues in southern California and parts of Australia.

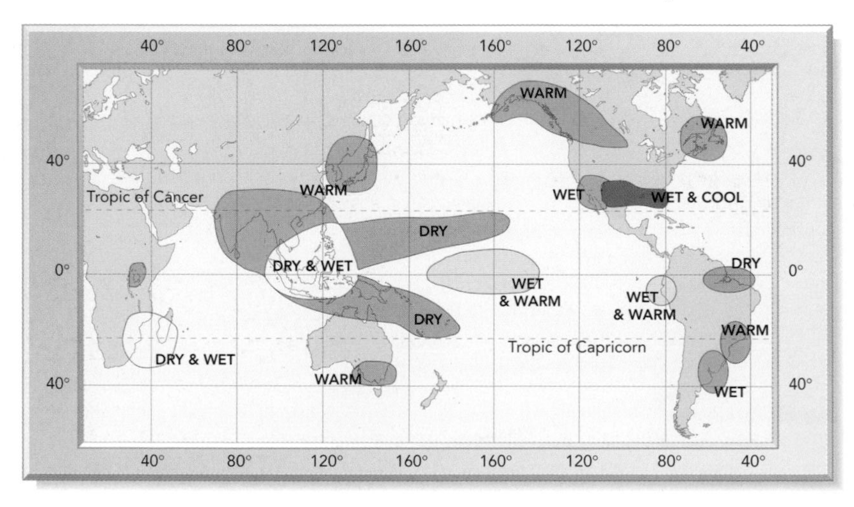

Typical thermal and rainfall anomalies (December to February) during a strong El Niño

chapter eight 8

Holiday Weather around the World

What we consider good holiday weather can vary from person to person. Most of us want dry conditions no matter where we are, the majority preferring warmth and sunshine, too. Those who enjoy outdoor pursuits are not always so fussy about high temperatures. Light winds with good visibility and broken cloud may be more to their liking. This chapter gives general information on the duration of various aspects of the climate, at a variety of holiday and other locations.

Monthly average values of weather data illustrate the seasonal changes that make up part of an area's climate. The variations from month to month in average temperature, precipitation, and sunshine are caused by the gradual evolution of the atmosphere's large-scale circulation patterns. The slow change in the intensity and location of the highs, lows, and ITCZ throughout the year control the nature and strength of the seasons at a particular site. When high pressure dominates, conditions will be dry, but they will be wet if low pressure, including the ITCZ, reigns.

Seasonal change

The perception of a season depends partly on whether a location lies within or beyond the tropics (from about 30°N to 30°S). In the tropics, which experience only minor temperature changes throughout the year, seasons are marked by the presence or absence of rainfall.

Outside the tropics, significant variation in the strength and duration of sunshine means that, in general, the seasons are defined on the basis of temperature change. There can be notable changes in precipitation, too.

Monthly values

Statements in this chapter about the duration of various aspects of the climate, at a variety of holiday and other locations, are quite broad in nature. The reason for this is that monthly values have been used rather than, for example, weekly averages.

The temperatures illustrated are average daily maxima and minima for a period of years. A given month's value – January, for example – is calculated by averaging that month's individual daily means of maximum and minimum temperature. If the location has a record of 30 years, the mean January values given are the averages of the 30 individual Januarys.

What constitutes a rainy day varies from place to place, so the amount is usually quoted in the description. Where local sunshine totals or other variables are not readily available, those from nearby sites with similar characteristics have been used.

St Petersburg, Russia

Although a coastal region, this part of north-west Russia is situated well away from very extensive areas of open sea. Therefore, it experiences quite a large annual range of temperature, from warm summer conditions – particularly when the air travels across the heated continent from a southerly quarter – to periods of intense winter cold under the influence of frigid air from either Eurasia or the Arctic Ocean to the north.

Winter is prolonged, the mean maximum only climbing above 0°C [32°F] between April and November. As in many middle- and higher-latitude locations, the temperature rises rapidly in the spring and falls dramatically in the fall.

Precipitation falls in the form of snow during the winter. It occurs typically on one out of two, or two out of three, days from November to March. The largest amounts of precipitation, however, are experienced during the summer and early fall. In the warmer months, showery rain falls as the continent is heated, while the occasional frontal system produces much of the total in the fall.

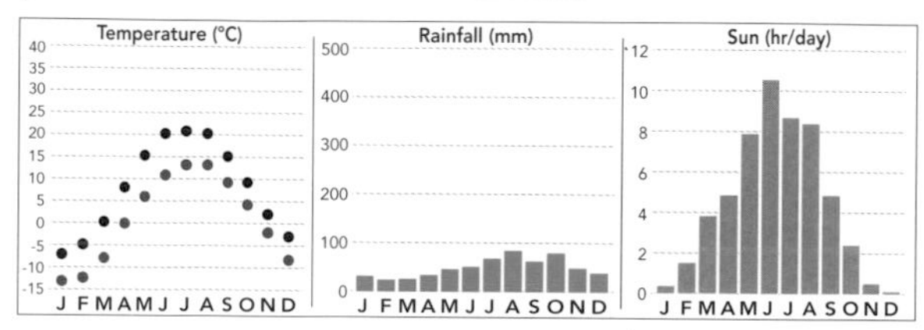

The annual amount of sunshine varies greatly at this high latitude. Conditions in the depth of winter are bleak, with an average of less than one hour of sunshine a day during November, December and January. The best month for sun is June, with over ten hours a day, while May to August generally sees more than eight hours.

Oban, Scotland

This site is reasonably typical of the western coast of Scotland, and is generally very much wetter than the eastern side of the country. Its temperature varies from an average maximum of about 17°C [63°F] in July and August to 6 or 7°C [43 or 45°F] between December and February. The mean minimum monthly value does not drop below zero in this maritime location, although frosts do occur.

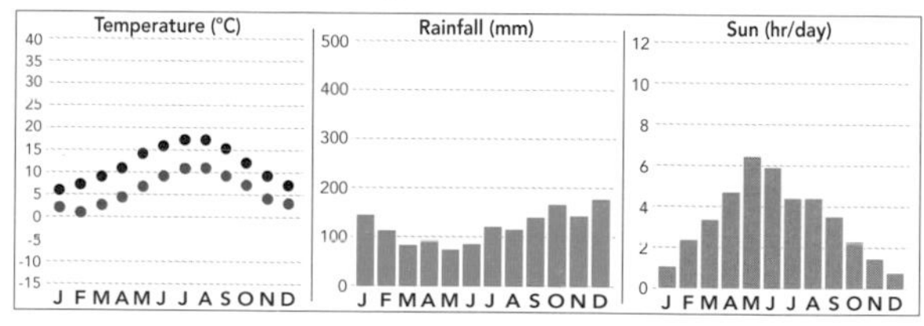

Rain is a feature of this region, since it is close to the tracks of the frontal cyclones and depressions that sweep across the North Atlantic, generally from south-west to north-east. Much of the rain is borne by these depressions; the driest period is normally during the spring, from March to June, when 0.25 mm [0.009 of an inch] or more of rain typically falls on one day in two. Sunshine in winter is very limited. It reaches its highest values in May and June.

Berne, Switzerland

Berne's temperature range is relatively large as befits a location some distance from the sea. Summers are warm, while winters are cold, the period from mid-November to mid-March seeing an average minimum below 0°C [32°F].

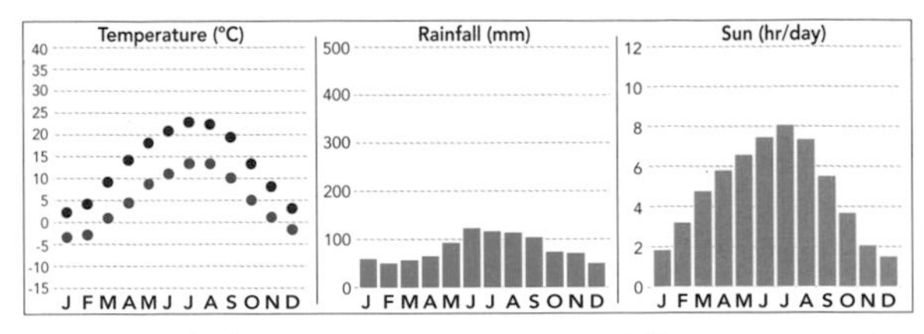

Much of the winter precipitation falls as snow, but the substantial seasonal maximum from May to September is usually rain associated with showery weather. This warm-season wetness is a common feature of continental interiors in middle and some higher latitudes. The pattern of rainy days, when 0.3 mm [0.01 of an inch] or more is observed, follows the same trend as the monthly totals. Sunshine totals are low in mid winter – generally two hours a day or less – while they reach a peak from June to August. July is usually the sunniest month, with an average of eight hours each day.

Athens, Greece

This location experiences a typical Mediterranean climate with hot, dry summers and mild, relatively wet winters. The hottest conditions are caused by outbreaks of intensely hot air from the Sahara. Maximum temperatures in the low 40s Celsius are not unknown during high summer.

Greece can suffer invasions of cold air from a northern quarter during the winter; in fact, in Athens, air frosts have been recorded on rare occasions between November and April. The northern part of Greece is more prone to this than the islands.

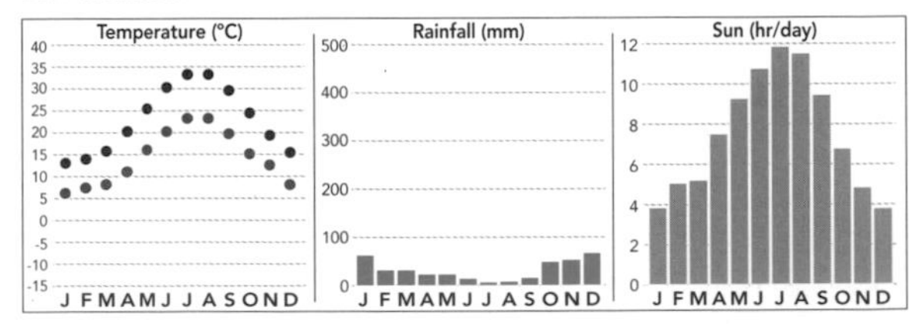

Rainfall is reasonably plentiful during the fall and winter; indeed, on average, Athens is wetter than London in December and January. The same is not true of the summer, when Athens experiences very small amounts of rain from

June to September. October is the month in the Mediterranean as a whole when the weather takes on a more disturbed state. The number of rainy days when 0.1 mm [0.003 of an inch] or more is observed varies from two or three a month, in high summer, to one day in two during December and February. Sunshine totals are at their best – more than ten hours a day – from June to the end of August.

Malaga, Spain

Another typical Mediterranean climate, but with a smaller annual temperature range than the more "continental" eastern part of the basin. This area's average summer maximum is 2–3°C [36–37°F] lower than that of Athens, while the winter maximum is 3–4°C [37–39°F] higher.

Malaga and south-east Spain generally experience a marked fall and winter rainfall peak, which is associated with low-pressure systems that can be frontal (often moving in from the Atlantic) or "cut-off" features. The latter are slow-moving, non-frontal low centers that can produce torrential rainfall. The number of days when 0.1 mm [0.003 of an inch] or more of rain falls ranges from an average of about one day in four, in the winter, to one or less a month in high summer. The higher total of rainy days experienced

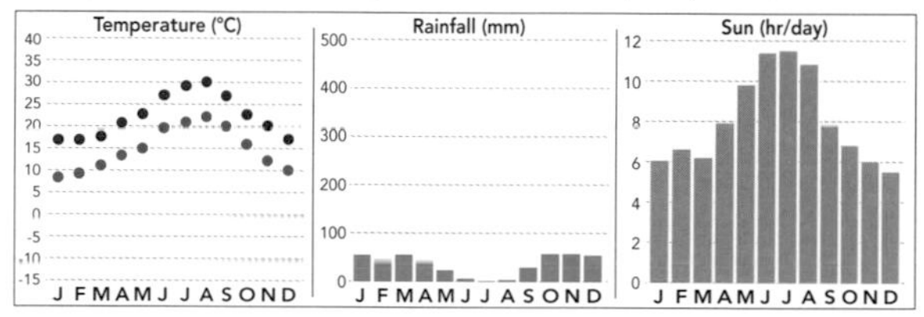

in the eastern Mediterranean region, exemplified by Athens, is due in part to the rain-bearing, low-pressure disturbances that actually form over the Mediterranean Sea and track eastwards. Sunshine is most plentiful from May to August, with July being the marginal winner. Even during the winter, the sunshine total is a respectable five or six hours a day. This means that a typical Malagan February is nearly as sunny as London in July!

London, United Kingdom

This characterizes the south-east of England, with a moderate annual range of temperature and, perhaps contrary to popular belief, quite moderate amounts of rain. The warmest months are July and August, which have only very marginally different mean temperatures. The hottest conditions in this area are associated with prolonged, cloud-free flow from a southerly quarter off the nearby continent, the more so if this has experienced a short sea track, which minimizes the cooling effect of the Channel on the air.

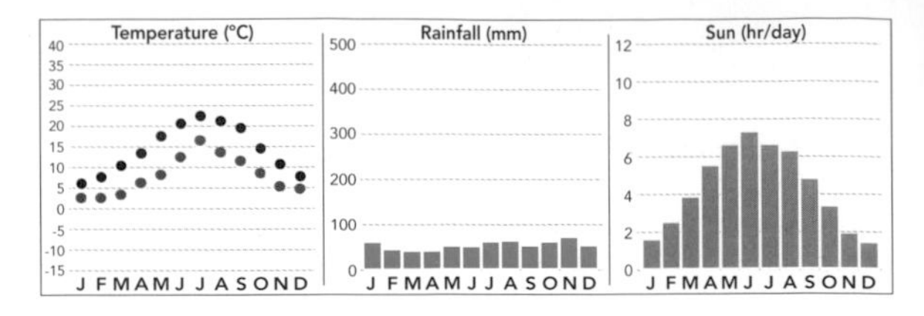

Snow is not very common in this large urban area and rarely settles. Rain is well spread through the year and tends to be showery in the summer, in contrast to the more widespread frontal type during the fall and winter. The annual fall of some 600 mm [24 inches] is typical of lower-lying south-east England as a whole. The number of days on which 0.25 mm [0.009 of an inch] or more falls varies from around one in two, during the winter, to one in three in the summer. Sunshine hours are low in the short winter days, and are best during June, although the period from May to August sees over six hours a day on average.

Sousse, Tunisia

This location is typical of the African coast of the western Mediterranean, having hot, dry conditions in the summer months in association with the northward movement of the subtropical anticyclone. The deep, sinking motion of the high provides prolonged sunshine, high temperatures and arid conditions, which characterize summer in much of the Mediterranean.

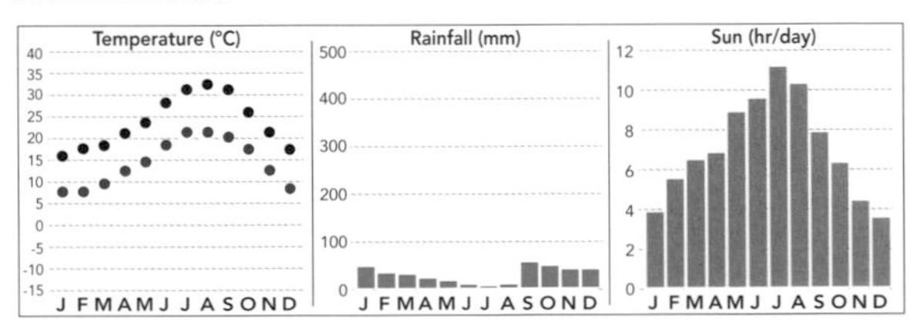

The marked increase in rainfall in the autumn and the persistence of fairly high totals (above 30–40 mm [1–2 inches] a month) are linked to the passage of frontal depressions across the region. Sometimes, these penetrate from the Atlantic, otherwise they form over the warm Mediterranean or in the lee (east) of the Atlas Mountains.

The wetter months experience rain on only a few days; in fact, the number of days when more than 0.1 mm [0.003 of an inch] falls ranges from an average of one in July to a maximum of nine in December. This means that even in the wettest months, rain occurs on only one day in three or four. Sunshine totals from Annaba, some 250 km [160 miles] distant on the Algerian coast, indicate a maximum in July of 11 hours a day. Over eight hours a day are experienced from around May to September, but this falls to four hours or so during November, December, and January.

Las Palmas, Canary Islands

These islands lie in a subtropical region that is dominated by the extensive Azores High as it intensifies and shifts northwards during the summer. Because the region is maritime, temperatures do not often reach uncomfortable extremes. Any really hot conditions are usually associated with a flow of air from the baking hot Sahara to the south-east, when summer temperatures can reach the mid- to high- 30s Celsius. In the main, winter temperatures are very comfortable, although cool air can sweep across the islands from the Atlantic to the north and north-west. Under such conditions, the minimum can fall to around 8°C [46°F] in extreme cases during December to March.

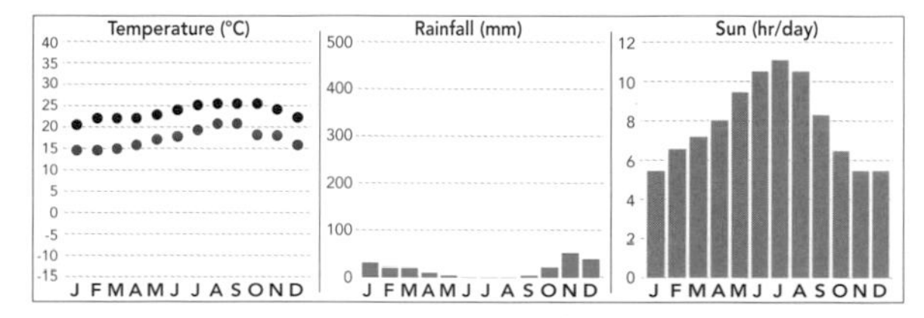

The area is very dry in the summer, particularly from May to September, during which an average of 20 mm [0.8 of an inch] falls on five rainy days (those experiencing 0.1 mm [0.003 of an inch] or more). During October to March, the region is influenced more by mid latitude disturbances, with an occasional frontal system or a slow-moving, low pressure 'cut-off', the latter characterized by scattered heavy showers. Sunshine totals (for Santa Cruz de Tenerife – broadly similar to Las Palmas) are highest from May to August, with a peak of 11 hours a day on average in July. The duration during the winter season is typically five or six hours.

Cairo, Egypt

This location experiences a very arid desert climate typical of a large part of north-east Africa. Cairo's weather is dominated by the generally cloudless skies associated with the normal presence of high pressure over the region. Heat is intense during the summer, the period from November to

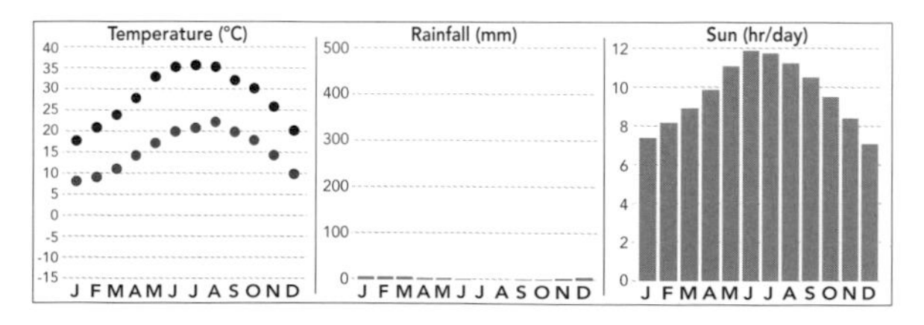

May being more comfortable in this respect. In the depth of winter, the minimum temperature can very occasionally fall to just above freezing in the cool desert nights.

Rainfall is extremely low. More than 1 mm [0.03 of an inch] falls on about six days in a typical year, and none at all is observed during June to August. The wettest months are December to March, with some 5 mm [0.2 of an inch] each. A good deal of this rain is produced by cold fronts that track over this part of Africa from winter depressions running across the eastern Mediterranean. Not surprisingly, sunshine totals in one of the world's major deserts are high. There is a seasonal variation in the duration, from seven or eight hours in the cooler season to over 11 hours a day from May to August.

Banjul, The Gambia

The small temperature variation between the warmest and coolest months in Banjul is fairly typical of a tropical location. The slight reduction in the maximum during July and August, which coincides with the higher minimum, is linked to the presence of cloud-laden and generally moist air from the winter (southern) hemisphere. This combination restricts daytime heating a little, while heat loss at night is suppressed.

The ITCZ influences this part of West Africa during the height of the summer monsoon, with very large rainfall totals producing over 1 m [3 ft] during July to September. During this period, measurable rain falls on slightly over one day in two. The single wet season is typical of a tropical location that lies at a latitude far enough from the equator to be influenced only once a year by the deep convective cloud and heavy rain of the ITCZ.

The quick decline to lower rainfall totals between November and May, during which precipitation occurs on average on only three days, is a result of the establishment of an extensive continental high across North Africa. These months also experience the hottest conditions during the day, and the coolest at night – mainly because this area is frequently influenced by cloud-free (although often dust-laden), dry air blowing from the Sahara to the east. The generally clear skies promote heat loss at night, which is unlike conditions during the summer monsoon season.

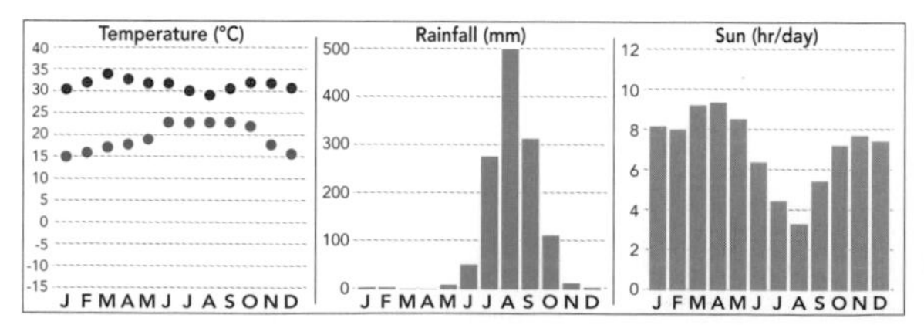

Mombasa, Kenya

A near-equatorial site where the annual range of temperature is very small. The coolest conditions occur in July and August, when the Kenyan and adjacent countries' coasts are affected by air flowing northwards, across the equator, from the winter (southern) hemisphere towards the low centered over the Indian subcontinent. The warmest months of January to March are also the driest, with an occasional shower on one day in ten in February, and one in five in January and March.

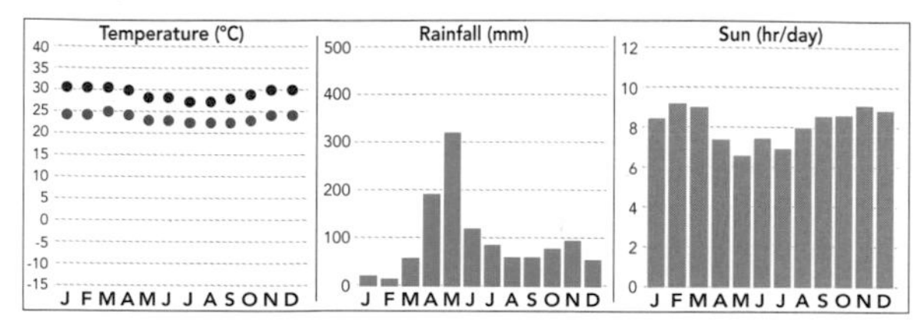

Mombasa's location means that it experiences two wet seasons associated with the gradual southward passage of the ITCZ across the equator during October and November (rain on one day in three), and the northbound crossing from April to July (rain on one day in two). These are known as the "short" and "long" rains respectively. Sunshine values follow a pattern broadly related to that of rainfall. The totals are relatively high all year round, but the most prolonged sunshine tends to occur during the driest, least cloudy period, in February and March. The lowest durations (although still around 6.5–7.5 hours a day) are observed from April to June, during the major rainy season.

Port Victoria, Seychelles

This island group lies in the western Indian Ocean, at the same latitude as Kenya. The very small seasonal change in average temperatures is an expression of its oceanic location; the ranges are a mere 3°C and 1°C [37 and 34°F] for the maxima and minima respectively. Extremely hot conditions are rare because the islands are relatively far from the continental sources of intensely heated air.

Rain falls during every month of the year, although it tends to take the form of very heavy showers. The driest months are typically July and August, when the ITCZ is at its northernmost point. During this period, rain falls on one day in four typically, and temperatures are slightly cooler, as the islands are influenced by the cross-equator monsoon flow from the southern hemisphere.

The wettest months of December and January see a total of some 700 mm [28 inches], which accumulates in falls on one day in two. This is the time when the ITCZ has moved south into the southern summer and is most active in the region. The north–south migration of the ITCZ is much more apparent over the tropical continents than the oceans, because they are subjected to massive changes in heating compared to the oceans. This means that low-latitude land locations often experience two precipitation seasons in a year, whereas oceanic sites tend to see one prolonged period that peaks in one maximum because the ITCZ does not migrate so extensively over the tropical oceans.

Further south, toward Mauritius, Madagascar and Mozambique, there is a risk of intense cyclones (the regional name for hurricanes) when the sea's surface temperatures are at their height. December to February is the main risk period for these disturbances.

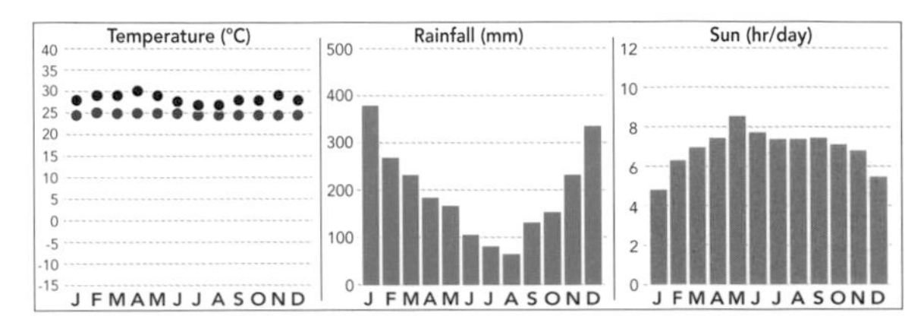

Harare, Zimbabwe

The elevation of this city means that temperature levels are normally very comfortable, the coolest conditions occurring from May to September. The warmest months are October and November, when the solar heating is increasing, before the arrival of the cloudier, wet conditions of the southward-moving ITCZ. These changes are reflected in the average sunshine totals, which are highest from August to October at more than nine hours a day typically. The least sunny period is within the rainy season. December, January, and February experience 6.1, 6.3, and 6.3 hours on average respectively.

Like Banjul near the northern limit of the ITCZ's seasonal movement, Harare experiences one rainfall season in summer, from November to March. One rainy day (with more than 1 mm [0.03 of an inch] observed) in three is common in November, rising to one in every two days in January. The dry season, from April to October, is dominated by higher pressure. During the first and last months of this period, rain falls typically on one day in six or seven – from May to September, this decreases to only one or two days a month.

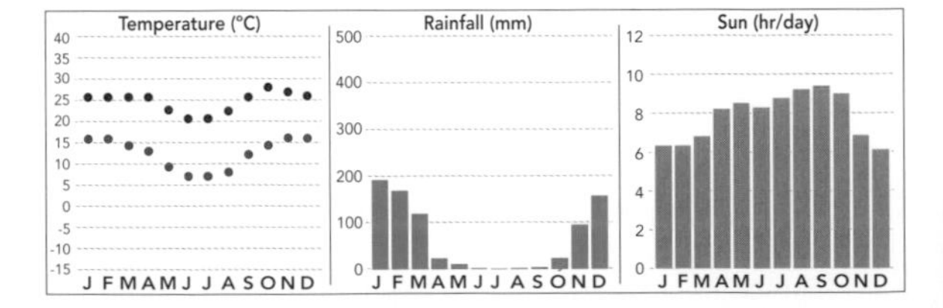

Durban, South Africa

Durban and the south-eastern coastal stretches of South Africa experience similar temperature levels to the Mediterranean region. However, the similarity stops there: the warm season is also the wet season, as the area is influenced by heavy showers from disturbances that develop in the interior and the south-west Indian Ocean.

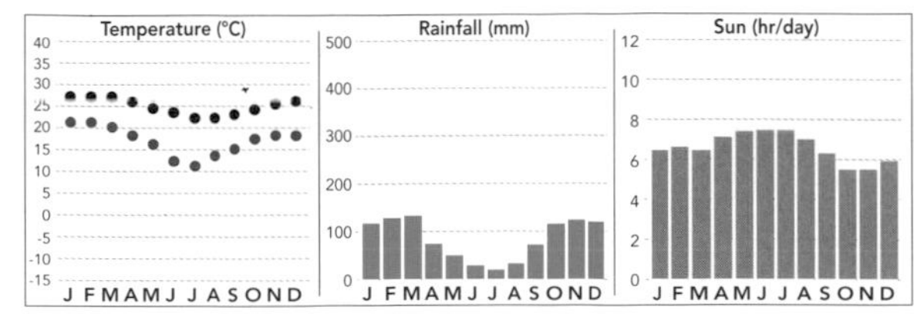

The warmest months are those of the southern summer, when the coastal areas can be influenced by very warm air from the interior. The driest period lasts from May to August, when on average more than 1 mm [0.03 of an inch] of rain falls on on only one day in ten and the region is influenced by the wintertime higher pressure of the interior. Because the warmest season is the wettest and, thus, the cloudiest, sunshine totals tend to be lowest then. The sunniest conditions occur normally from May to July.

Cape Town, South Africa

Unlike Durban and the adjacent coastal area of South Africa, the Cape Town region has a Mediterranean-like climate. The summer is warm; the winter mild. Rainfall occurs on average during every month, but the main season is the winter, when this region is strongly affected by the frontal depressions that sweep from west to east across the South Atlantic Ocean.

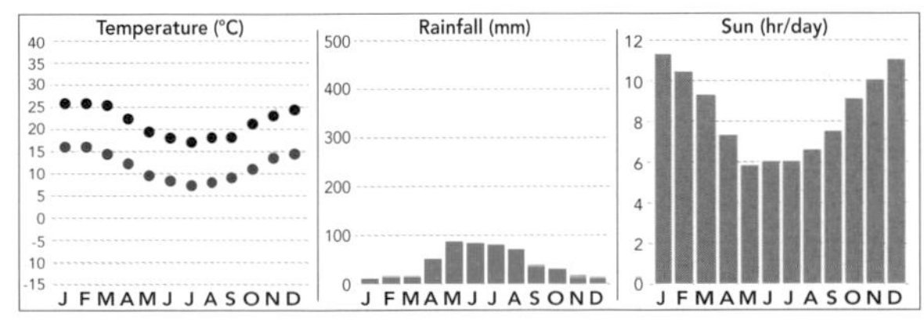

Normally, the wettest conditions are encountered from May to August, when the number of rainy days (those with 1 mm [0.03 of an inch] or more) is about one in three. This decreases to one day in ten or 15 between November and March. The area is quite sunny: totals in the summer are 10–11 hours a day, and in winter around six or more.

Boston, Massachusetts, USA

Although it is a coastal site, Boston – and most of the north-east coastal area of the USA – experiences a large annual range of temperature between the warmest months of June to August and the cold winter conditions from December to March. The former are associated with the common flow of warm and very moist air from the south-west in the summer, around the western flank of the Azores High. The latter are caused by the outbreak of cold air from the high pressure that normally occurs across the interior of North America during the winter.

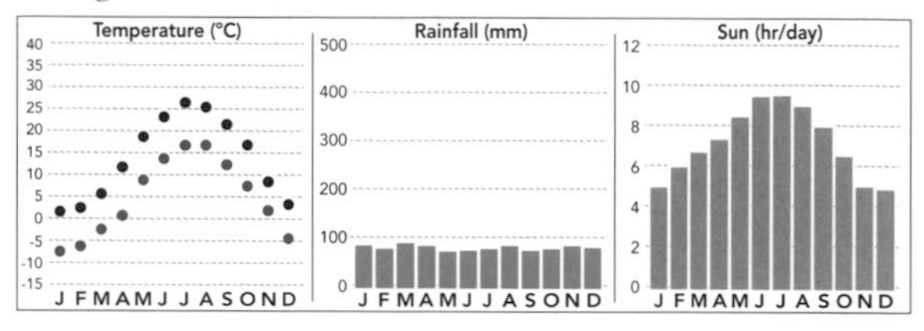

Precipitation is well distributed throughout the year, each month averaging a value between 79 mm [3 inches] in May and 97 mm [4 inches] in March. The number of days when more than 1 mm of precipitation (or about 10 mm [half an inch] of snow) falls ranges from six in October to nine in the following five months. The risk of snow is highest during December to February. Boston enjoys the sunniest conditions between May and August.

Grand Junction, Colorado, USA

This high-level site, deep within the continental interior, is broadly typical of many central Rocky Mountain locations. It experiences a large annual range of temperature between the hot conditions of high summer and the cold of the deep winter months. Mean monthly minima at, or below, 0°C [32°F] occur from November to March inclusive. Sunshine totals vary from five or six hours a day in the worst months to higher values between May and September. June and July experience more than 11 hours of sunshine a day.

Precipitation in the coldest season is in the form of snow, with higher totals in the Rockies around this area. As a rule, 1 mm [0.03 of an inch] of water is equivalent to 1 cm [0.4 of an inch] of snow. This site is dry, only some 200 mm [8 inches] a year falls on an average of 46 days (more than 1 mm [0.03 of an inch]). More rain tends to occur from August to October, in part because of deep convective storms.

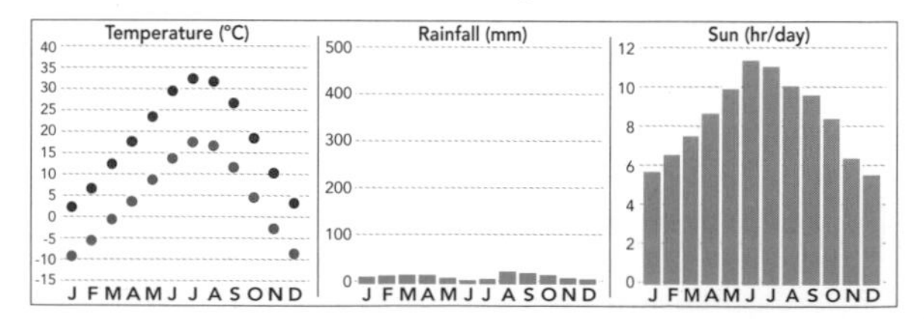

San Francisco, California, USA

For its latitude, the central Californian coastal area is cool, but temperatures are very much higher not far inland, away from the influence of the coastal fog and stratus cloud. Temperatures near the coast reach their highest and lowest in late summer and early fall. The winter is quite mild.

The thermal conditions in this region are strongly, but indirectly, influenced by the cool California Current, which flows south (toward the equator) along the coast. Moist air blown over this water is often cooled sufficiently to reach its dewpoint temperature – the classic sea, or advection, fogs that lap over the Golden Gate Bridge are expressions of this. Apart from the fog, stratus cloud is also common along the coast. During the warmer months, it burns off so that just a little inland, temperatures become very much higher.

Rainfall is quite strongly seasonal, indicating the dominance of oceanic high pressure in the summer and traveling frontal depressions during the winter. In that sense, San Francisco's rainfall regime is quite like that of a Mediterranean climate. The dry season of three or fewer rainy days a month (when more than 1 mm [0.03 of an inch] falls) runs from May to October inclusive; July and August are

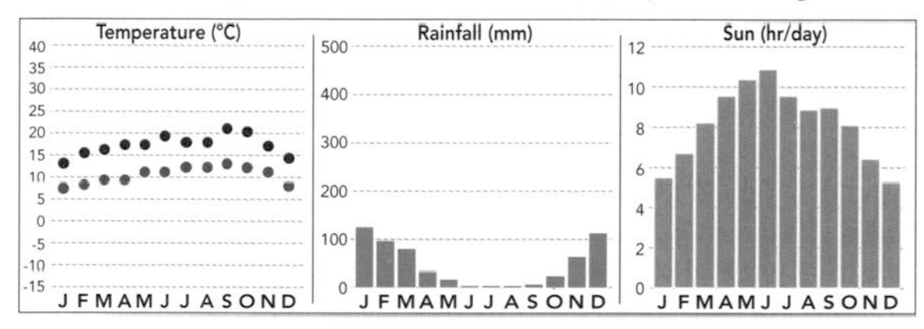

commonly completely dry. The wet season stretches from around November to March, when all months see substantial totals from falls that typically occur on one day in three or four. The sunshine totals reach their highest in May and June, although the whole period from April to September sees average daily durations of nine hours or more.

Jacksonville, Florida, USA

This area typifies broadly the changes that occur throughout the year across lower-lying regions of the south-east United States, with hot and humid conditions in the summer and a maximum of rainfall during the same season. Frosts occur occasionally when this area is under the influence of the cold air from the wintertime interior of the United States. Sunshine totals reach their peak, on average, in April and May. Although June and July are also sunny, their slightly lower duration values mark the start of a decline toward September.

This pattern is an expression of the increased risk of tropical, cloud-laden weather disturbances in this region as a whole, including the occasional tropical cyclone that may develop into a hurricane. These cyclones are most frequent in the tropical North Atlantic Ocean during September, although they are quite often seen from July to October. Cyclones vary in intensity from a depression, through a storm, to a full-blown hurricane; the first, weakest category is the most common.

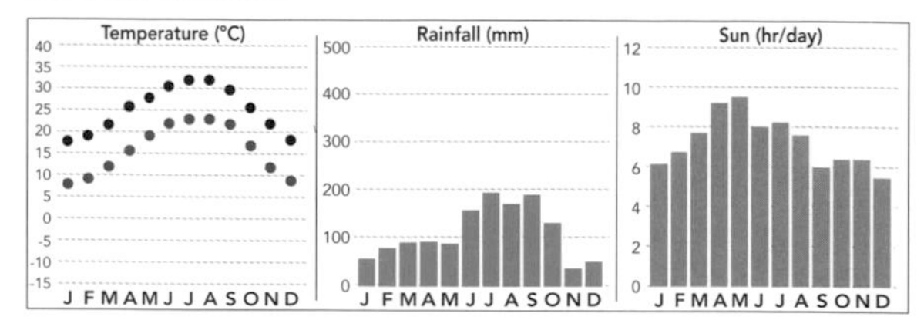

The driest season is the period from November to March, during which 1 mm [0.03 of an inch] or more of rain occurs on something like one day in five or six. The much wetter season, when intense, but often short-lived, rain occurs, peaks during June to September. Then one in three, or even one in two, days see rain.

Barbados, Windward Islands

This is a maritime site, and broadly representative of many of the Lesser Antillean islands. The region is influenced year-round by the Trade Winds of the tropical North Atlantic, which tend to have more cloudy disturbances embedded within them when the sea temperature is high.

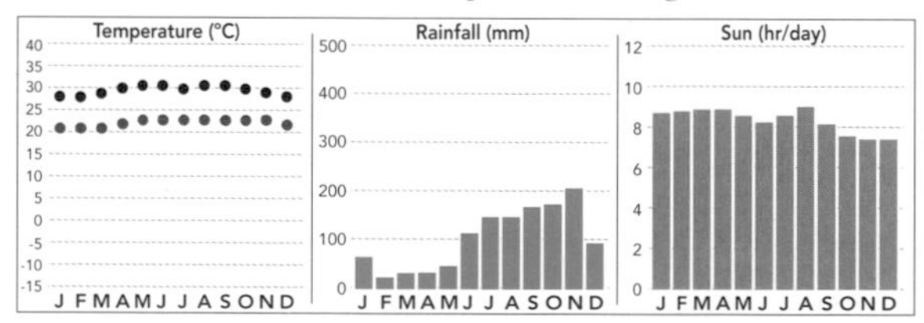

The annual range of temperature is typically very low: the warmest months see up to 31°C [88°F], while the coolest experience a very warm 28°C [82°F]. Like many tropical sites, Barbados' seasonality is based on the strong pattern of rainfall that develops during the year. Thus, the drier period dominates the months of February and March. It is during this time that rain (more than 0.25 mm [0.009 of an inch]) falls on about one day in three, four or five. The wetter season peaks during September to November, when the total falls are high and typically occur on one day in two. Generally, except when the island experiences a slow-moving tropical cyclone, rainfall is intense and short-lived. Sunshine totals are good all year round, although there is a tendency for slightly lower durations from September to December.

Buenos Aires, Argentina

This city represents conditions in the low-lying region of central eastern Argentina and neighboring Uruguay. It experiences quite a wide temperature range for a coastal site. This is because it is susceptible on the one hand to summertime outbreaks of stiflingly hot, humid air from Amazonia, and on the other to relatively cold air from the south-west and south in the winter. Very broadly, its temperature regime through the year is similar to that of the Mediterranean.

The number of rainy days (when more than 0.25 mm [0.009 of an inch] of rainfall occurs), is fairly constant throughout the year, with marginally higher frequencies from July to December, and generally lower frequencies during the remaining months. Rainfall totals vary to a greater extent, larger amounts being observed from October to

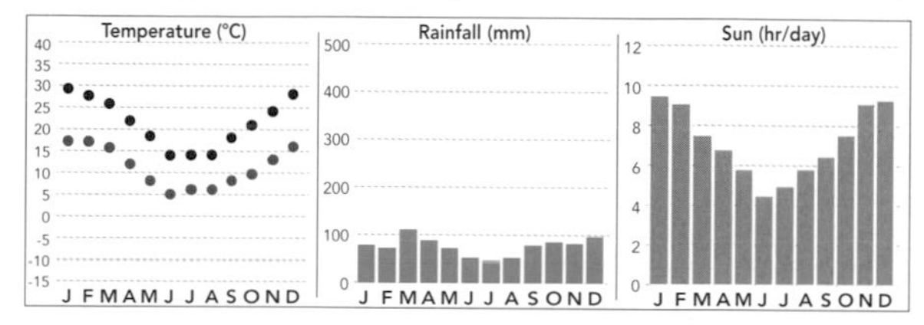

December and during March and April. Drier conditions are seen normally during the winter, from June to August. In the warm season, heavier rain often falls as intense bursts from deep cumulus clouds. During the winter, the rain tends to come more from the mid-latitude frontal depressions that track across the region from the direction of the Andes.

Sunshine hours are pleasantly high in the summer, but fall by about 50% during the winter months. Overall, the sunshine values are similar to those of the Mediterranean.

Cairns, Queensland, Australia

Weatherwise, the northern part of Australia is a truly tropical region, and Cairns is no exception. There is a relatively small annual range of temperature between the warmest months of December to February and the marginally cooler ones of June and July. Extremely high temperatures would be associated with a flow of bakingly hot summer air from the Australian interior to the south-east.

As with many tropical locations, rainfall marks the seasonal change. Here, the wettest conditions occur during December to April, with very wet weather in January, February and March. This marked maximum is associated with the penetration of the ITCZ into parts of northern Australia during their summer season. Moreover, the Queensland coast – and northern Australia in general – is prone to tropical cyclones during the season of warmest sea temperatures. This is December to March in the southern hemisphere. The average number of days when 0.25 mm [0.009 of an inch] or more rain falls ranges from about two in three during March to one in four from June to October.

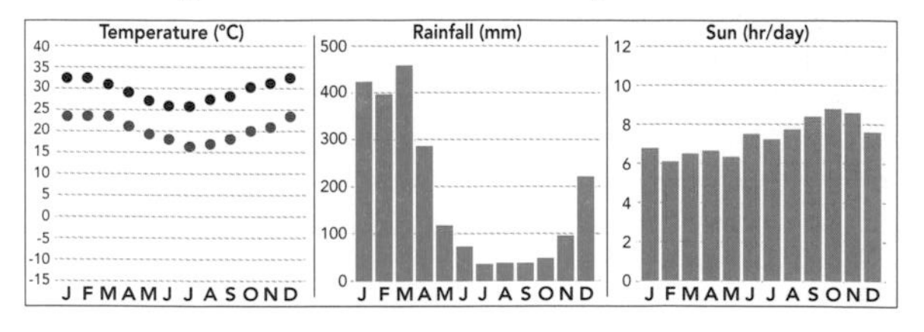

Sydney, New South Wales, Australia

The south-eastern coastal sector and the southern fringe of Australia are, very broadly, similar to the Mediterranean in climate. Both regions are prone to very hot conditions from extensive deserts toward the equator, and to winter cold conditions that are ameliorated by tracts of sea towards the poles. Sydney is protected from cold Antarctic blasts by the Southern Ocean, while North African coastal sites are similarly shielded from cold European air by the less-extensive, but warmer, Mediterranean.

During the winter, this area experiences rain from frontal depressions that brush across southern Australia in the westerlies of the Southern Ocean. Much of the southern coastal region of this continent has such a winter rainy spell. Rain falls during the summer, too, but more often in intense bursts from deep convective clouds. The number of days during which 0.25 mm [0.009 of an inch] or more is observed does not vary much during the year – from about one day in three in August to one day in two during January. Sunshine totals range from the highest during October to December to the poorest between April and June.

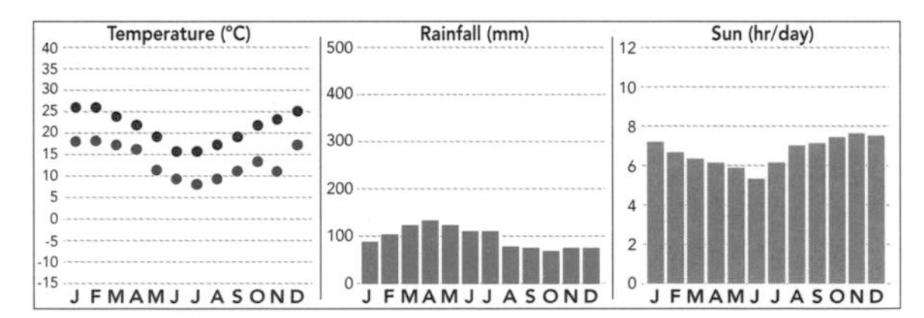

Auckland, New Zealand

This area is essentially maritime, although it is not as "conservative" in its seasonal temperature variation as similar locations well within the tropics. The oceanic influence has ensured that during a 100-year period, the highest temperature ever recorded was 32.2°C [90°F] in January and

February, while the lowest was 0.6°C [33°F] in July. New Zealand is far from extensive continents that act as sources of hot air in the summer, and cold in the winter. Any outbreaks of such air from Australia or Antarctica respectively are considerably ameliorated as they cross extensive oceans.

The warmest months are January and February, while the coolest is July. Sunshine totals are best from November to February, and generally poorest from May to July.

Fairly large falls of rain occur during most months, with a pronounced maximum during the winter, when many active frontal depressions sweep across New Zealand from the west. The four wettest months are May to August inclusive, when 0.25 mm [0.009 of an inch] or more falls during two in three days on average. Tropical disturbances do influence this region in the summer and tend to produce heavy, more short-lived rainfall. Consequently, during the summer months, rain occurs typically on one day in three.

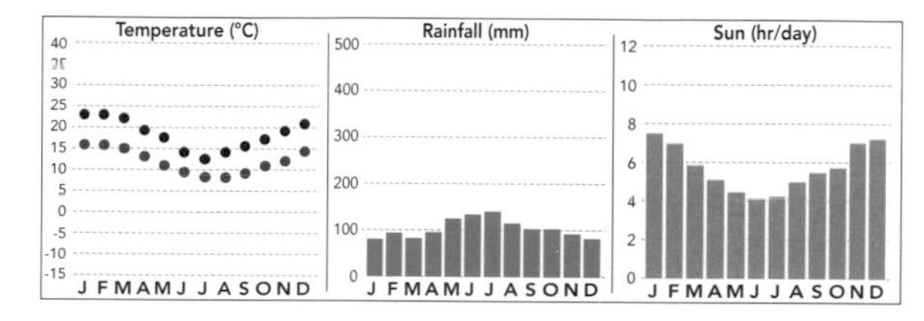

Bahrain, The Gulf

The Gulf, is almost completely enclosed and is a very warm sea. The region is one of the most arid in the world, lying below the persistent deeply subsiding air of the subtropical highs. This means that many days of the year are characterized by unbroken sunshine and, in the summer in particular, by unbearably hot conditions during the day. There is a seasonal variation in temperature, making the winter more comfortable than other times of the year.

The highest temperatures are associated with flow from the interior of Saudi Arabia in the summer, while the lowest come from the direction of Iran and the cold Asian continent in winter. Air frosts are very rare, but have been recorded, on occasion, far inside Saudi Arabia during the depth of winter.

Not surprisingly, rainfall is extremely sparse: the number of days in an average year when 2.5 mm [0.09 of an inch] or more are observed is a mere nine. Nevertheless, there is a seasonality to the rain, the major part being recorded from November to March. Wet days in this period are usually due to the passage of cloudy troughs that sometimes move through the Gulf from the eastern Mediterranean. They are sources of rain for many areas of the Middle East. However, there is often no rain whatsoever in Bahrain – and the Gulf in general – from June to October inclusive. Sunshine in the Gulf region normally is most prolonged from May to October, with 10–11 hours a day. This falls to seven or eight during the coolest months.

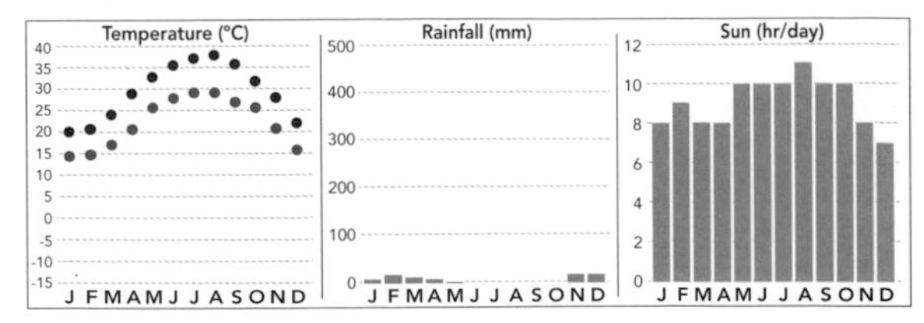

Hong Kong, China

This city lies in a region of Asia that is susceptible to cooler air from the Asian interior in winter. The highest temperatures occur during the wet season, when conditions are classically "hot and steamy". Such weather is most apparent from May to September, when the maxima and minima are elevated in conjunction with high rainfall totals due to the progress of the ITCZ across the region. Total rainfall is usually above 2 m [7 ft] a year, which comes mostly in dramatic, intense bursts from tropical thunderstorms. There is a risk of even more dramatic amounts of rainfall over a day or two from typhoons. These are the cousins of hurricanes; in fact, they are the same kind of system with a different regional name. Like hurricanes, they grow out of features that begin as tropical depressions, then become tropical storms. They are most common in September on average but can occur during other months too.

These tropical cyclones (large-scale low-pressure disturbances, ranging in intensity from depression to typhoon) are much more frequent in the north-west Pacific than any other tropical ocean region. In fact, some 33% of all the world's disturbances of this sort occur there, partly because it is unusually warm compared to the other ocean regions.

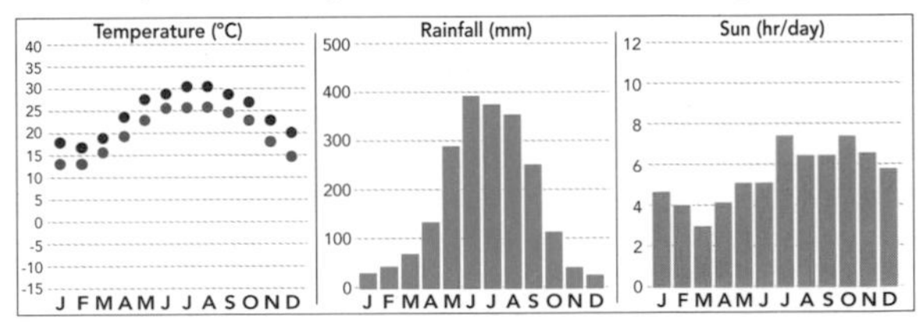

The height of the rainy summer monsoon season sees precipitation on one day in two, while the frequency during the dry winter monsoon season drops to between two and five days a month from November to February. During the latter period, the area is often dominated by the extensive Asian winter anticyclone. The occasional outbreak of cool air from this high can lead to air temperatures approaching 0°C [32°F] in extreme cases. Sunshine totals are best towards the end of the rainy season, when values typically reach a peak in October and November. The cloudiest conditions are usually experienced from February to April.

Colombo, Sri Lanka

This island enjoys a classic low-latitude climate, with a very small annual range of temperature and double rainy seasons associated with the north- and southward passages of the ITCZ. The hottest months tend to be in the late spring and early summer, before the cooler, cloudier air from the winter (southern) hemisphere penetrates the region as the summer monsoon sets in.

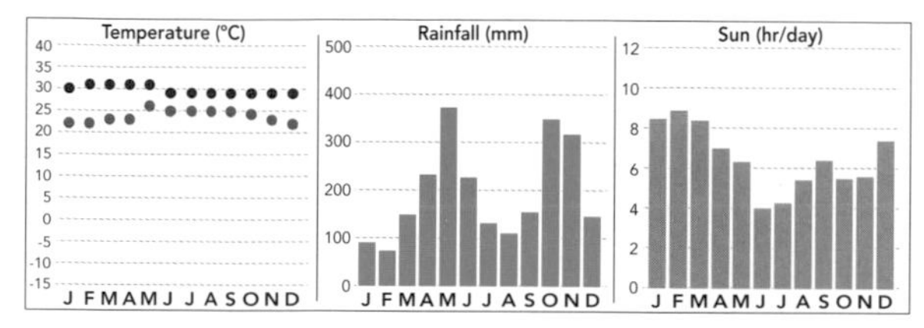

Every month of the year sees rain, although there is a strong double peak within the annual round. The first occurs from April to June, as the ITCZ moves slowly north toward India, and the second during October and November, as it crosses the island on its southbound migration. The great towering cumulonimbus clouds in this region of convergence produce short, but intense, rain showers. Sometimes the clouds merge to become larger-scale, longer-lived disturbances that generate prolonged heavy rain.

Colombo's annual fall of over 2 m [7 ft] is fairly typical of a monsoon location at sea level. Rain falls on about two days in every three during the height of the ITCZ's presence, decreasing to about one in four or five days from January to March. This is when the region is most strongly affected by the less-disturbed flow coming from Asia in the winter.

Sunshine at a broadly similar location, Trivandrum, some 300 km [190 miles] to the north-west on the Indian coast, indicates that the best totals occur during the winter monsoon from December to March, when skies are often cloud-free. There is a fall in sunshine to a minimum at the very time of year when the potential is highest – in June. This is because of the frequent occurrence of cloud and rain in the summer monsoon. There is also a second decline during the southward passage of the ITCZ in October and November.

Phuket, Thailand

This site is typical of a low-latitude location near sea level, having persistently hot and humid conditions. The highest temperatures tend to occur, as with many other monsoon-influenced sites, before the onset of the rains associated with the migrating ITCZ. Thus, the slightly hotter period is normally during February, March, and April. The number of rainy days increases sharply from only six in March to 19 in May, while the respective monthly average totals range from 74–297 mm [3–12 inches]. There is a marginal decline in the rainfall around June, before the second, larger increase occurs during September and October. The dry season stretches from December to March. The driest weather of all occurs in January and February. Sunshine totals (from Songkla, some 250 km [160 miles] to the south-east of Phuket) indicate that the best conditions occur from January to April, with upward of eight hours a day. Once the prolonged wet season has set in, however, the duration falls to a little over five hours daily from September to November.

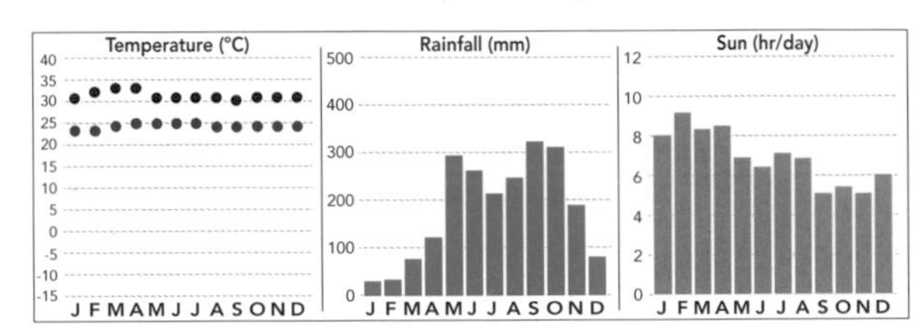

chapter nine

Environmental Issues

During the last decade or so, we have become more aware of global, or near-global, changes in the atmosphere that have occurred naturally or through the activities of man. Perhaps the most well known of all these issues is the phenomenon known as El Niño. El Niño, global warming, and the "hole" in the ozone layer are the three environmental issues that are described in detail in this chapter.

Despite recent extensive news coverage, El Niño is nothing new – it has probably occurred intermittently for thousands of years. When it rages, it has a very significant widespread impact over a few seasons. Scientists are gradually unraveling its secrets, using improved monitoring and advanced modelling techniques.

Global warming has also occurred in the past, even before the existence of the human race; the climate naturally fluctuated over time. However, scientists are quite certain that the increase of greenhouse gases in the atmosphere will lead to a significant rise in surface air temperature in the coming decades. This increase in such gas concentration is almost certainly due to the large-scale industrial and agricultural activities of man.

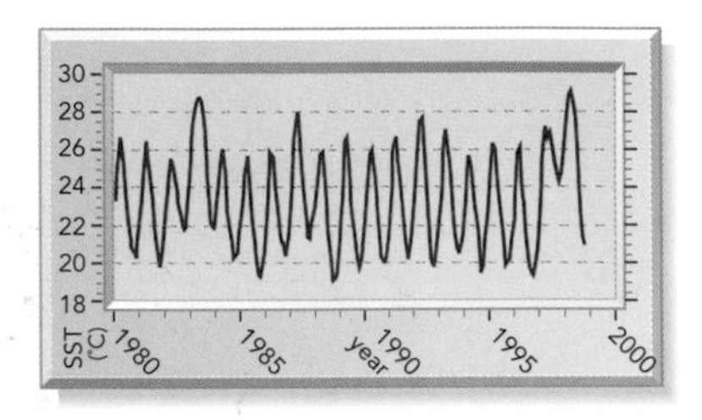

Sea surface temperature west of Peru and Ecuador 1980–98

The so-called "hole" in the ozone layer is undoubtedly another problem caused by human activity. It is more correctly known as Antarctic stratospheric ozone depletion. The discovery, some 15 years ago, of the plunge in ozone concentrations during the southern-hemisphere spring, stimulated rapid and effective international action to ban the gases that scientists had shown to be the culprits.

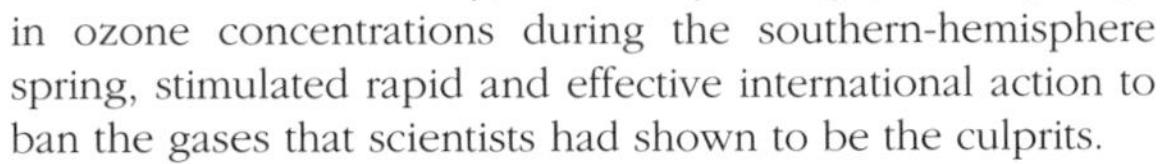

These three environmental issues will be described in detail in this chapter.

El Niño

The term "El Niño" is Spanish for "the little boy", and its use to label a significant ocean-warming event in the coastal region of Peru and Ecuador dates from a few centuries ago. The earliest written evidence of probable El Niño-related weather in South America comes from Spanish colonists who settled in these regions during the late 15th century.

The name was coined by fishermen whose livelihood was affected by the annual appearance of a warm ocean current, typically around Christmas time – hence "El Niño", or "the Christ Child". The current persisted in the region for a few months, and one outcome was that fish – anchovies, for example – were much less abundant during this period. The reason for this was that they fed in the nutrient-rich cool water that was usually present offshore. They were a staple ingredient of the local economy, which suffered as a result.

Wider significance

Over the years, the term "El Niño" has been applied to the very much larger-scale and much more intense ocean warming that occurs occasionally across huge stretches of the equatorial Pacific. Today, the marked warming near Peru

World Meteorological Organization
http://www.wmo.ch

and Ecuador, which was noted all those years ago, is known to be only a relatively local signature of a Pacific-wide change. In fact, on occasion, El Niño leads to very substantial thermal and rainfall anomalies in areas far from the Pacific Ocean.

The Southern Oscillation

We know that El Niño is associated with an atmospheric phenomenon known as the Southern Oscillation (SO). This term was coined by Sir Gilbert Walker in 1923, when assigned to the Indian Meteorological Department. In an effort to find a means of predicting the nature of the Asian summer monsoon, he studied very large-scale pressure variations and discovered that "when pressure is high in the Pacific Ocean, it tends to be low in the Indian Ocean from Africa to Australia".

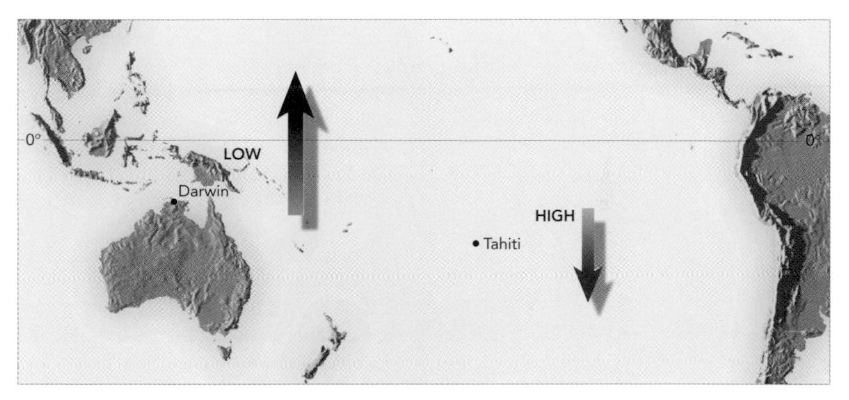

Simultaneous rise/fall in surface pressure at Darwin, Australia and Tahiti

Pressure is normally high over the southeast Pacific and low in the western equatorial Pacific. The horizontal gradient of pressure between these two centers leads to the presence of the easterly (westward blowing) Trade Winds. During such times, the SO is said to be in its High Index. Sometimes, however, the barometer falls over a period of some months across the eastern Pacific and, when this happens, it rises simultaneously in the western Pacific.

This change leads to a weakening of the pressure gradient, together with a weakening – or even a reversal – of the Trades. This is known as the Low Index of the SO. Walker noted that during such phases, there were droughts in Australia, Indonesia, India, and parts of Africa. In addition, winters tended to be unusually warm in western Canada. Furthermore, the desert islands in the middle equatorial Pacific suffered persistent and torrential rains.

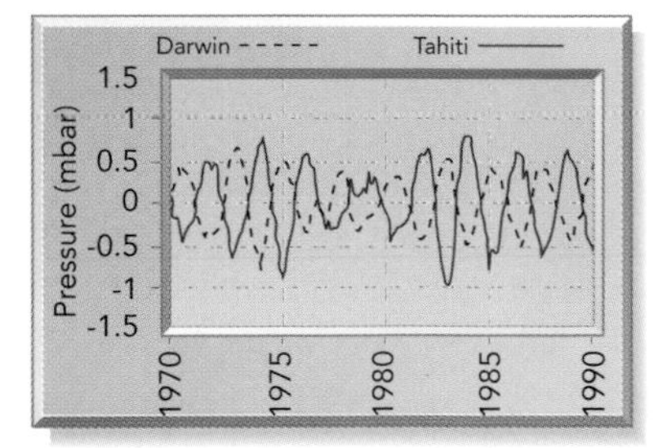

Surface pressure anomalies for Darwin, Australia, and Tahiti 1970–90

ENSO

In the late 1960s, Professor Jacob Bjerknes was the first to suggest a possible link between the SO phenomenon and El Niño. He proposed that the unusually warm sea-surface temperatures were probably related to the weak, or even reversed, easterlies and the torrential rain of the Low Index phase. He showed that the SO and El Niño are both parts of the same phenomenon, known collectively as ENSO. Today, an SO index is quoted, being the difference in monthly mean-sea-level pressure between Tahiti, representing the South Pacific anticyclone, and Darwin, representing the western Pacific low.

Normal conditions

During normal, or average years, the easterly winds that blow along the equator, and the south-easterlies that blow along the coast of Ecuador and Peru, drag surface water along with them. The rotation of the Earth deflects this water to the right of the flow in the northern hemisphere, and to the left in the southern. This means that water is driven away from the equator in both hemispheres and also from the Peru/Ecuador coastline. Cold, nutrient-rich water wells up from below to replace it, forming narrow zones of equatorial and coastal upwelling less than 150 km [100 miles] wide.

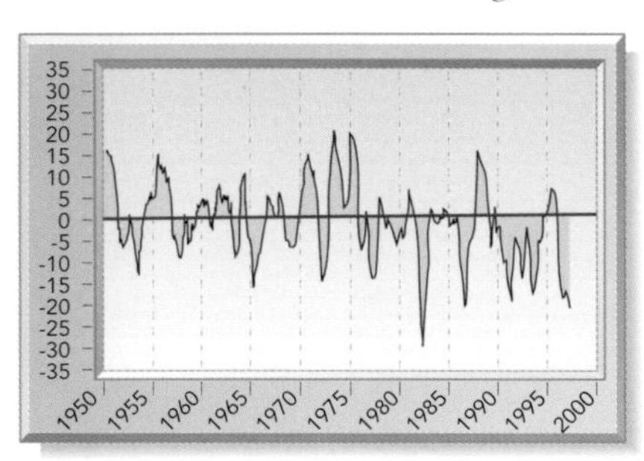

Smoothed values of the Southern Oscillation Index. Negative values indicate warm ENSO phases

Normal conditions are marked by cool temperatures over the eastern Pacific and a warm maximum over the equatorial western Pacific. This western area is so warm that very deep convective cloud and heavy rainfall are a hallmark. Part of the huge volume of air that ascends to great height within these clouds – as high as the upper troposphere, in fact – moves eastwards at these levels and sinks in depth across the eastern Pacific. This vertical circulation is called a Walker cell, after Sir Gilbert. A number of these cells exist around the equator, connecting wet and dry regions. The descending portions of such cells are characterized by very dry and often cloud-free weather.

Thermocline

Within the ocean, there is a layer with a depth of some 100 m [328 ft], through which the water temperature drops rapidly. Known as the thermocline, it separates the upper warmer zone from the much colder deeper reaches. Normally, the thermocline is near the surface in the eastern equatorial Pacific, some 50 m [160 ft] down, and it slopes gently down toward the western side, where it is found at a depth of about 200 m [650 ft].

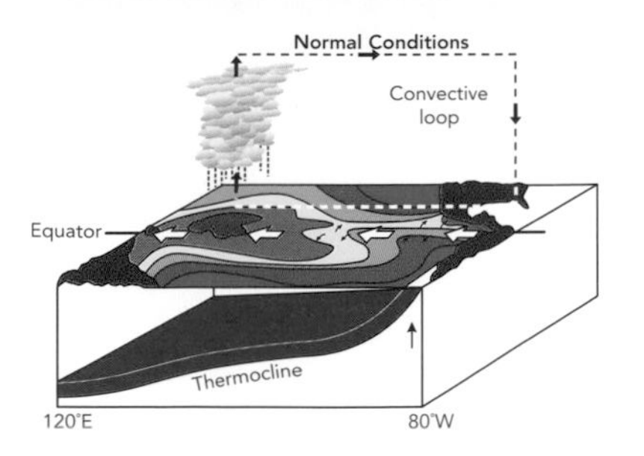

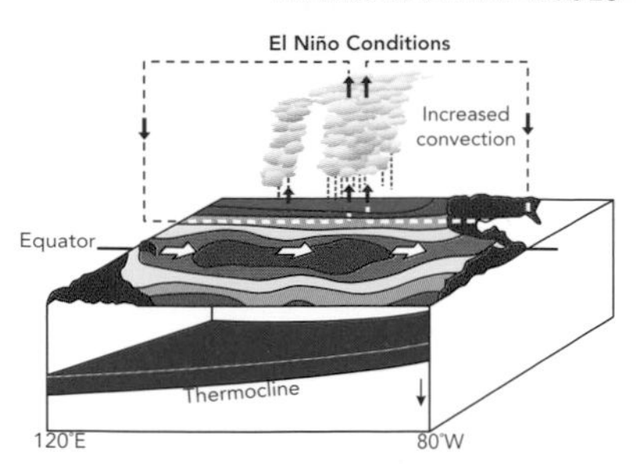

Normal (left) and El Niño (right) conditions, including sea surface temperature, thermocline depth, and Walker cell location

If there were no wind stress on the ocean's surface, the thermocline would be more-or-less horizontal. In this region, however, the persistent Trades drive water westwards, lifting the thermocline toward the surface in the east, and depressing it in the west. The fact that the westward-driven surface water is steadily warmed by sunshine and, therefore, is of lower density, means that the sea's surface slopes up toward the western equatorial Pacific. When the Trades are blowing at their strongest, the sea level in the western basin is over 0.5 m [2 ft] higher than in the east. This very broad, flattened mound of warm water occurs around Indonesia and New Guinea.

As the SO index gradually moves to a low-value phase, when the Tahiti/Darwin difference is small, the relaxation in the normally strong Trades leads to the thermocline becoming less tilted. It drops by more than 100 m [328 ft] in the east and cuts off the cool upwelled water from the Ecuador/Peru coastal zone.

Wave across the Pacific

The sea level flattens out along the equator, falling in the west and rising in the east. In association with this, the warm surface water flows eastwards as a long, low wave known as a Kelvin wave, reaching South America a few months later, where it turns north and south along the coast. This leads to an increase in sea level and the migration of fish. The northward branch of warm water influences marine life as far north as Vancouver, Canada.

The eastward migration of the warm water across the equatorial Pacific causes the air above it to become moist and warm. It gains sufficient buoyancy to produce massive convective cloud and torrential rain in regions that otherwise are persistently arid. The classic El Niño conditions are for normal rainfall across the central equatorial ocean, and unusually dry conditions over the western sector, including eastern Australia, Indonesia and over northern South America. This distinct pattern is due to the fact that one portion of the enormous upward movement of air in the displaced convective region subsides over the western Pacific.

After the Kelvin wave has left the western Pacific, the warm water layer thins substantially and mixes with cooler water. This cooling leads to less evaporation and a more stable atmosphere, which together mean less rain. The eastern half of Australia and all of Indonesia, therefore, are susceptible to drought during a marked El Niño.

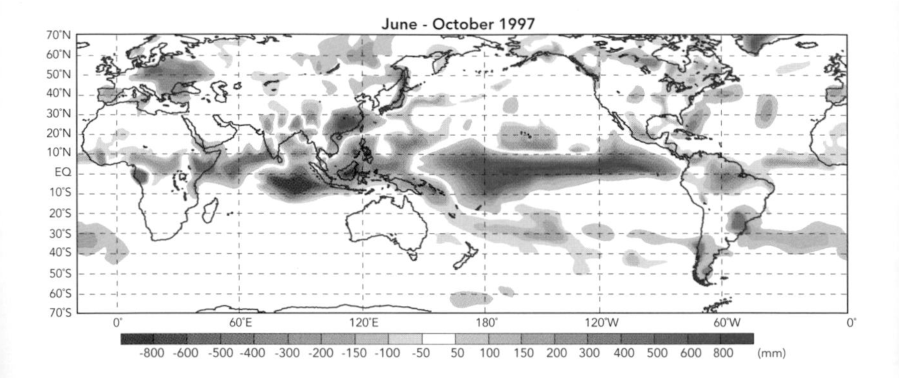

Estimated rainfall anomalies (mm) for June to October 1997. Wetter (green), drier (brown).

Unusual patterns

There are typical thermal and precipitation anomalies associated with a strong El Niño. The northern hemisphere summer (June to August) that follows the evolution of the El Niño tends to be characterized by drier than average conditions in the region containing Indonesia, Australia, and the Fijian Islands. Drought is a risk across the northern part of South America and the southern Caribbean, too – and there is evidence that the Indian monsoon may be drier. The extensive forest fires that affected large areas of Indonesia and Malaysia in September 1997 were related, in part, to the unusually dry El Niño-influenced preceding months and stable weather conditions around that time. Wet conditions occur in the central equatorial Pacific because of the significantly higher sea-surface temperature.

The northern hemisphere winter (December to February), like the summer, illustrates the major tropical pattern that is linked to shifts in location and intensity of the Walker cells. Dry conditions stretch from Sumatra and southern Malaysia to the Hawaiian Islands in the north, and the Fijian Islands in the south. Additionally, drought is a high risk for eastern equatorial South America and south-eastern Africa.

Abnormally wet conditions occur over Ecuador, Peru, southern Brazil, Uruguay, northern Argentina, southern USA, and equatorial east Africa. Warmer than average winter conditions are often experienced from Alaska to the Canadian Rockies, in parts of southeastern Canada and northeastern USA, and around Japan.

How does the effect spread?

We have seen how El Niño can influence the weather patterns within the tropics, but how does it produce wet weather in San Francisco, Tampa, or Buenos Aires? The answer lies in the impact that the unusually located, deep tropical rain clouds have on the upper atmosphere.

The deep clouds that move across to the central equatorial Pacific pump very large amounts of heat and air high up into the troposphere, over a very extensive region. Shifting the thunderstorms from the western Pacific to a point thousands of kilometers further east, strengthens the upper tropospheric flow significantly.

The change in the strength and pattern of the winds at 200 hPa can be highlighted by comparing the average picture for January with that of January 1998. The strength of the jet has increased notably across both the North Pacific and southeast USA. There are more subtle, but nevertheless significant, changes in the wind direction depicted by the thin streamlines with arrowheads. The more vigorous, wavy jet means that there is a likelihood of more active frontal systems (wetter and windier) that may be marginally, but significantly, displaced from their normal location.

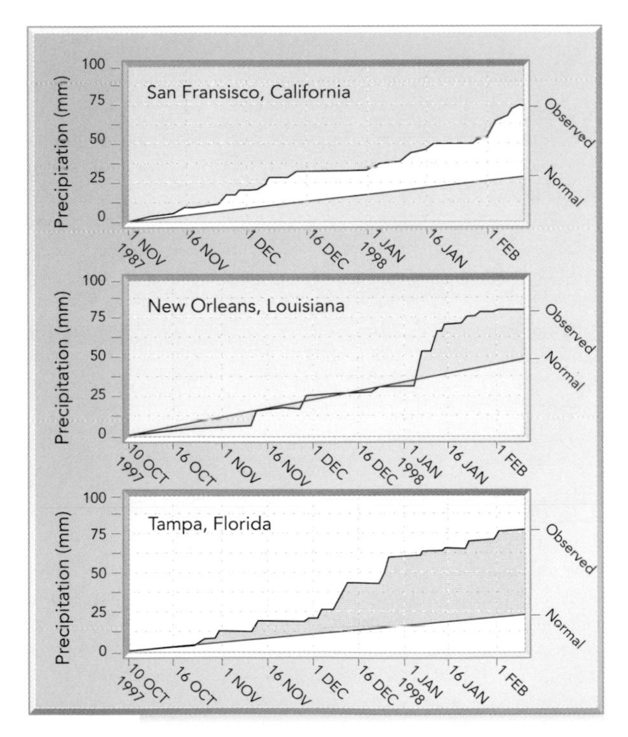

Observed (October 1997 to early February 1998) and average accumulated rainfall for three US cities

The cost

Marked El Niños do not occur regularly. Within the last two decades, the world has seen two dramatically influential examples of this phenomenon – in 1982–83 and 1997–98. Because they are associated with extreme regional weather events that persist for a season or so, they are costly. The 1982–83 El Niño was estimated to have caused extensive damage – the resulting costs shown below:

Flooding	
US Gulf States	$1270 million
Ecuador/Northern Peru	$650 million
Bolivia	$300 million
Cuba	$170 million
Hurricanes	
Hawaii	$230 million
Tahiti	$50 million
Drought/fires	
Australia	$2500 million
Southern Africa	$1000 million
Mexico/Central America	$600 million
Indonesia	$500 million
Philippines	$450 million
Southern Peru/Western Bolivia	$240 million
Southern India/Sri Lanka	$150 million

A total bill of US$8110 million! It is very likely that the 1997–98 event will cost more.

Predicting the event

Because an El Niño grows slowly over months and seasons, and can be monitored from weather satellites and a special surface network of observation sites across the equatorial Pacific, timely warnings of the risk of abnormal conditions can be issued well ahead of the event. The characteristic anomalies that we are aware of form one basis for such warnings. In addition, climate modeling and prediction centers use sophisticated computer models to issue predictions of the likely outbreak, evolution, and decline of an El Niño.

La Niña

The equatorial Pacific not only experiences El Niño however; on occasion it can experience its antithesis known as "La Niña", or "the baby girl".

During La Niña the central and eastern tropical Pacific waters tend to become much cooler than average. The recent El Niño decayed rapidly during January to April 1998

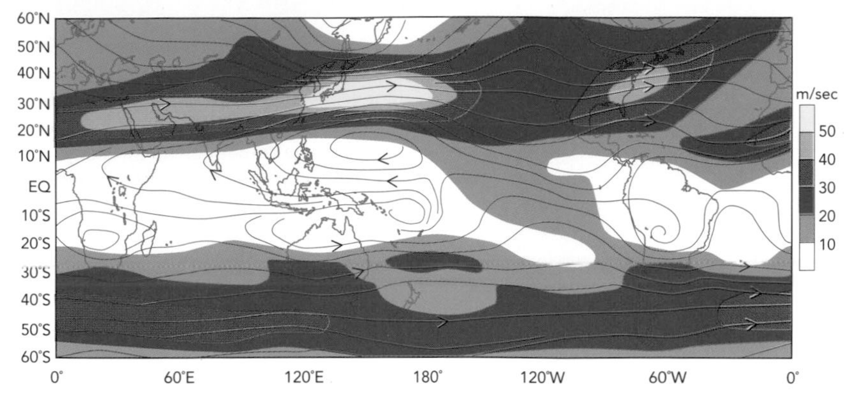

December–February mean 1979–95

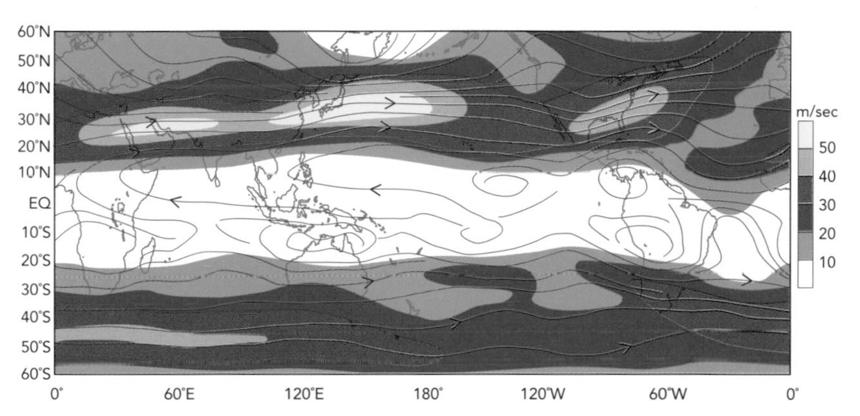

December 1997–February 1998 mean

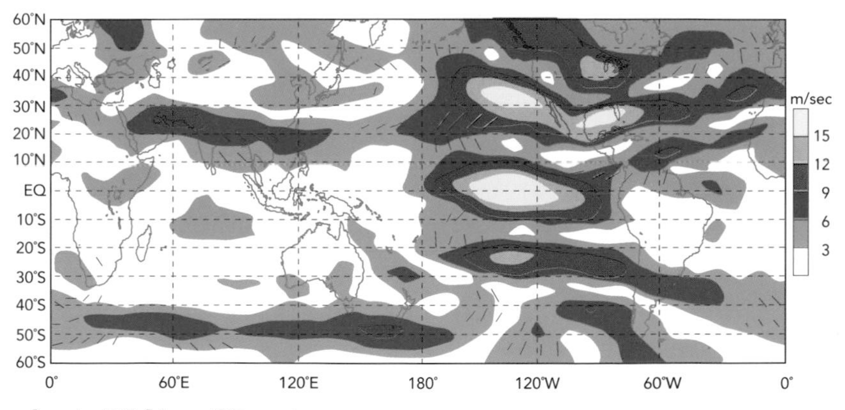

December 1997–February 1998 anomaly

250 mbar winds (streamlines and isotachs)

and in May and June the sea surface thermal anomaly in the central equatorial Pacific changed from one degree Celsius above to one below the long-term average. La Niña's cold ocean pattern then developed during the second half of 1998.

As this evolved, large-scale rainfall patterns were changing in a way we would expect them to under such conditions. Relatively wet weather occurred across large areas of Indonesia, Australia, and southern Africa while lower than average rainfall was observed over southern Brazil, Uruguay, northern Argentina, and east Africa.

La Niña is also linked to generally cooler than average surface land temperatures across the tropics and subtropics. There was evidence of this emerging in very early 1999.

It should be noted that there is also evidence of increased tropical storm activity in the North Atlantic during La Niña and decreased activity during El Niño.

Global warming

Greenhouse gases have been an integral part of the atmosphere for many millions of years. Their significance lies in their ability to absorb outgoing terrestrial long wave radiation and re-radiate in all directions, including back down to the surface. This means that greenhouse gases act as crucially important insulators for the Earth and all life upon it – without them, we would not be here.

The mark of society

Our industrialized society has been increasing the atmospheric concentration of these significant gases for the past century or more. We know from careful monitoring and estimation that carbon dioxide levels in the atmosphere have increased in the last 200 years, from around 200 parts

Earth surface losing heat to space

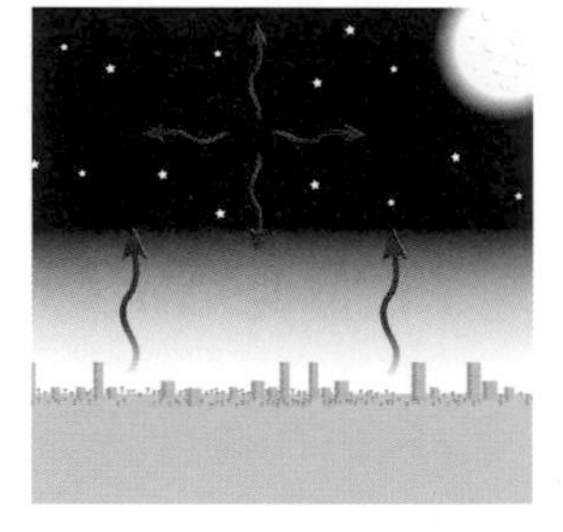

Greenhouse effect: some gases absorb part of the outgoing heat and re-radiate part of it back down

per million (ppm) to about 360 ppm in the late 1990s. Additionally, the concentration of atmospheric methane has doubled during the last century, while nitrous oxide is increasing at about 0.25% a year. These three gases are increasing largely as a result of energy generation, transport, and agriculture.

Because of the increase in these and other greenhouse gases, such as the halocarbons CFC-11 and HCFC-22, we can expect to see the impact of an enhanced greenhouse effect. Undoubtedly, human society is changing the composition of the atmosphere.

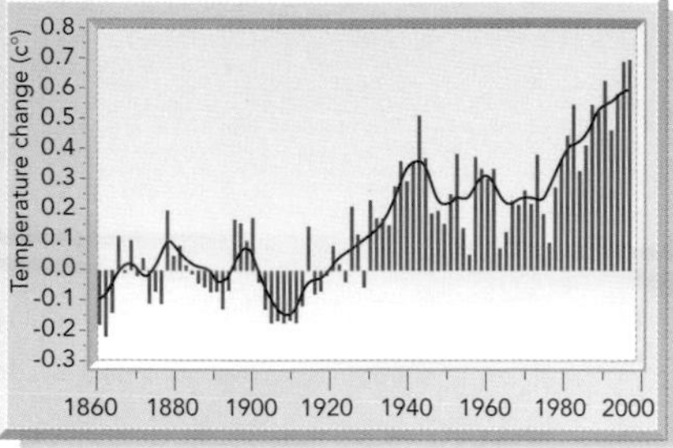

Departure of annual global air temperature over land and sea from 1961–90 average

The natural greenhouse effect is influenced primarily by water vapor and, to a much smaller extent, carbon dioxide. It is enhanced, however, by artificially increased levels of carbon dioxide, methane, CFCs, and nitrous oxide.

Water vapor is a powerful greenhouse gas, but at present its total global concentration is not changing significantly. If the surface warms as predicted, however, it would be expected to rise because of increased evaporation from the oceans.

Temperature change

So what has happened to the global average surface temperature over, say, the last century or so? The combined air temperature over land and sea-surface temperature for each year from 1861 to 1997, when compared to the average for the period 1961–90, is illustrated above. Each line above or below the 0.0 axis represents a positive or negative departure in respect of the recent 30-year period. The curve is a running mean that smooths out short-term fluctuatons, while the red curve depicts the increasing concentration of atmospheric carbon dioxide over the same period.

When averaged around the globe, it would seem that there has been an increase of 0.6°C [1°F] from start to finish during this particular period. It is important to stress that, although the carbon dioxide has increased markedly during this time, global mean temperature has not risen continuously. There were periods, during the 1940s and 1950s, for example, when it remained constant or even decreased.

Modeling the change

Climate modelers use sophisticated computer simulations to represent the complex physical interactions we know to be important in determining climate and its changes. The variables and the interactions that concern such groups are highlighted on the next page.

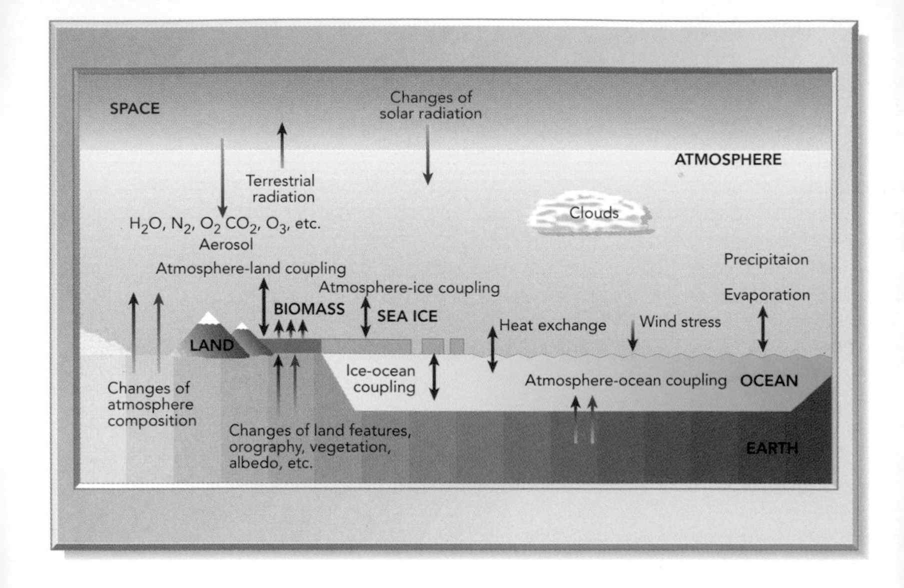

Climate system components and interactions

Where does the extra gas go?

The artificial increase in greenhouse gases is only part of the story. The amount of carbon dioxide, for example, that remains in the atmosphere depends on quite poorly understood aspects of the global carbon cycle. The oceans absorb and release it in certain areas, while the world's vegetation cover also plays an important role, since it is absorbed by plants in the process of photosynthesis.

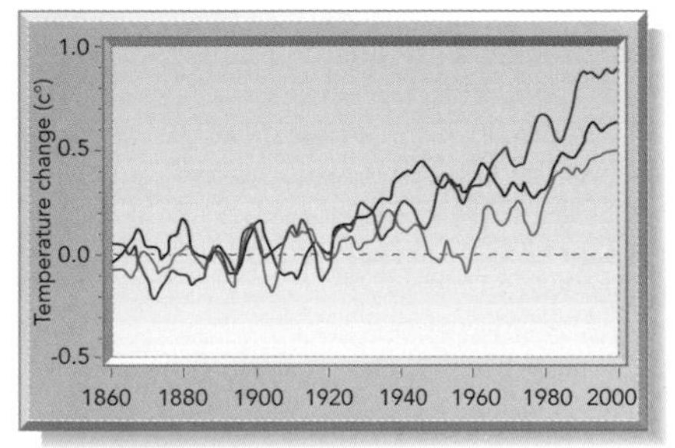

Observed (red) and simulated global temperature, including greenhouse gases (blue) plus sulfate aerosols (green)

The study of global budgets of methane, carbon, nitrogen, and other substances is still very much in its infancy. However, major international scientific projects have been developed to improve the observational networks, to stimulate research and, ultimately, to deepen our understanding of the reservoirs and the exchanges between them.

Aerosols

The impact of aerosols – very small particles suspended in the atmosphere – must also be considered by climate modelers. Recent work has highlighted the important role played by sulfate aerosols, which emanate from sulfur dioxide emissions and from the evaporation of sea water. They are known to offset the warming effect of greenhouse gases by causing the atmosphere to reflect more solar radiation into space than otherwise would be the case.

Climate models

Researchers need to know whether the climate models are any good at simulating the story so far. It is no good running a model forward for a number of decades if it does not do a reasonable job of reproducing the current climate and any past changes we know about. The UK's Hadley Center is renowned for this type of research. Depicted are the observed global temperature changes since 1860 in red; the computer simulation of change based only on increasing greenhouse gases in blue; and another simulation to which has been added the impact of sulfate aerosols in green. Of course, it would be miraculous to see a perfect, or even near perfect, fit.

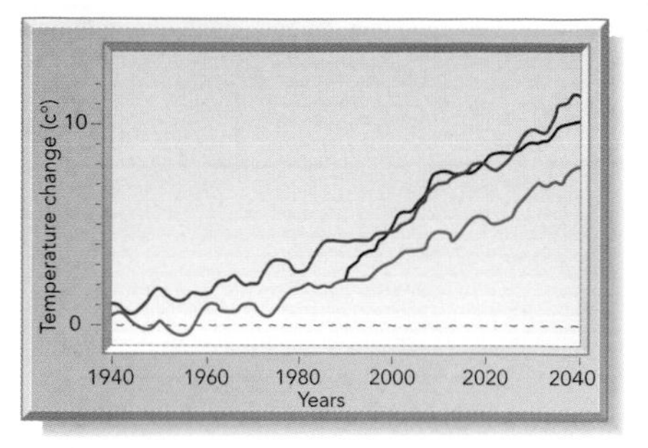

Predicted global mean temperature due to just greenhouse gases (blue) plus aerosols (green). The purple line depicts warming if aerosols cease in 2000

The simulations have, in essence, followed the fluctuations around the zero line from 1860 to about 1920 and, in the main, have tracked the broad increase since then. For the last 40 years or so of the series, the added-sulfate model has performed best, falling within 0.1°C [0.2°F] of the upward march of global temperature. This does not necessarily mean that the scientists understand the processes well enough to reproduce the observed change in this way, but it is unlikely that the reasonable match occurred by chance.

The best bet?

When the Hadley Center climate model is run forward in time, assuming an annual increase in atmospheric carbon dioxide of 1 ppm a year, it predicts a mean global warming of 0.3°C [0.5°F] a decade, the most intense increase occurring in the winter across northern high latitudes. The blue line on the graph above illustrates this progression, while the green line simulates the predicted change if sulfates are added – the increase then drops to around 0.2°C [0.4°F] a decade. An important difference in these two constituents is that aerosols remain in the atmosphere for several days, while carbon dioxide may linger in the atmosphere for up to a century.

If legislation reduced or leveled sulfur dioxide emissions, the model suggests that the full effect of warming due to carbon dioxide increase would soon become apparent. This is illustrated by the purple line.

The models are used also to predict the increase in global mean-sea-level that is expected to occur during the 21st century as the oceans warm and expand. The best estimate, due to this effect, is that the global average increase between now and the end of the 21st century will be 0.25–0.35 m [0.8–1.1 ft]. This is a very significant increase.

Uncertainties

The climate modelers themselves are the first to admit to the imperfections of their simulations, including the representation of clouds. Additionally, not enough is known about the sensitivity of climate to changes in aerosols, volcanic activity, and solar output.

Climate prediction work is carried out by a number of other groups, too, but all the models point in the same direction, with varying degrees of intensity and patterns of change. If the release of greenhouse gases into the atmosphere continues on a "business as usual" basis, it is estimated that carbon dioxide levels will double from pre-industrial values by the end of the 21st century. Adding the effect of increased water vapor and other factors leads to a spread of predictions about the magnitude of the warming by that time. They range from 1.5–4.5°C [35–40°F], the best estimate lying somewhere around 2.5°C [37°F] as a global average.

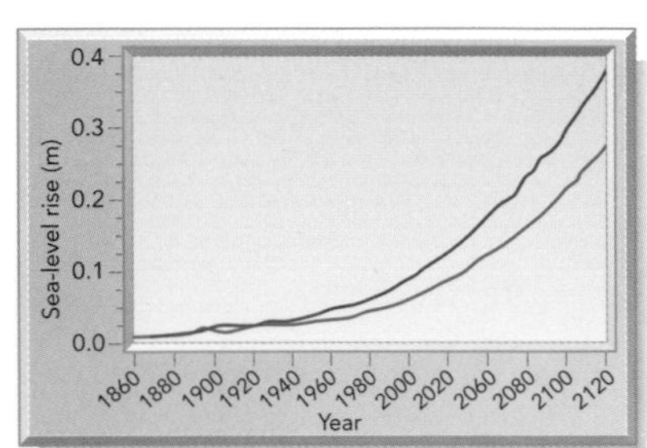

Global mean-sea-level increase due to expansion related only to greenhouse gas increase (blue) plus aerosol increase (green)

There are significant uncertainties in all this work, some of which have been mentioned. The subject is not completely cut and dried for all meteorologists. Some refute that an increase in atmospheric water vapor content will lead to a rise in greenhouse warming. They say that the increase will lead to changes in clouds, such that they will become more efficient at producing precipitation and, thus, raining out a lot of the water that would enhance the greenhouse effect. The real impact of increased water vapor will only become apparent some years into the future, when more modeling studies have been undertaken and more observational evidence is available.

One thing is apparent: an increase in water vapor could very likely lead to an increase in the incidence of cloud and, therefore, a decrease in the amount of solar radiation reaching the surface. This, of course, would be a negative feedback.

Overall, even though there are serious sceptics in the scientific community, and the models used to predict future atmospheric conditions are imperfect, a significant majority of atmospheric scientists place faith in the results provided by their peers in this particular field.

Decision time

Governments are listening closely to the predictions and are acting both individually and collectively to reduce the worst impacts of the predicted changes. There have been a number of very high-profile intergovernmental meetings to thrash out agreements on cutting back the output of greenhouse gases. At the 1992 Earth Summit in Rio de Janeiro, Brazil, it was agreed that by 2000, the emissions of carbon dioxide would be stabilized at 1990 levels to slow the rate of increase. To prevent further rises in its atmospheric concentration, a global 60% cut would be required. The likelihood of this happening is remote, to say the least.

Despite the Rio agreement, US carbon emissions are likely to surpass 1990 levels by some 10% by 2000. More generally, the World Energy Council has predicted that, by 2020, global carbon dioxide levels will have risen by between 42 and 93%, depending on the strength of economic growth. The Berlin Mandate of 1995 committed governments to negotiating a protocol at the Kyoto, Japan meeting in late 1997. The outcome of this gathering was not as positive as it might have been for reducing the intensity of global warming. The following round of talks was at Buenos Aires, Argentina, in late 1998.

Ozone depletion

What is ozone?

For many millions of years, the gas called ozone has existed naturally in the Earth's atmosphere and has helped to safeguard life as it has evolved on the planet. Oxygen is usually thought of in its diatomic form, O_2, which makes up slightly less than 21% of the atmosphere by volume. It is the second most common gas in the atmosphere after nitrogen (N_2) and is essential for the maintenance of life on our planet. Ozone, however, is the triatomic form of oxygen (O_3) produced by the bonding of a single atom (O) and O_2.

Natural creation and destruction

The solar radiation we see at the Earth's surface has been changed substantially on its way down through the atmosphere. Some of it is reflected back out to space by clouds, dust layers, and the land and ocean: reflected sunshine is completely lost to the Earth's atmospheric system. On the other hand, some of it is absorbed by the clouds and dust, and, of course, by the surface of the Earth. This absorption heats the substance involved.

Most of the solar radiation with wavelengths of up to 0.21 micrometers is absorbed above 50 km [32 miles] by nitrogen and oxygen. Absorption of incoming solar radiation in the range 0.21–0.31 micrometers is carried out principally by ozone within the stratosphere. Very short wave ultraviolet radiation, at wavelengths of less than 0.24 micrometers,

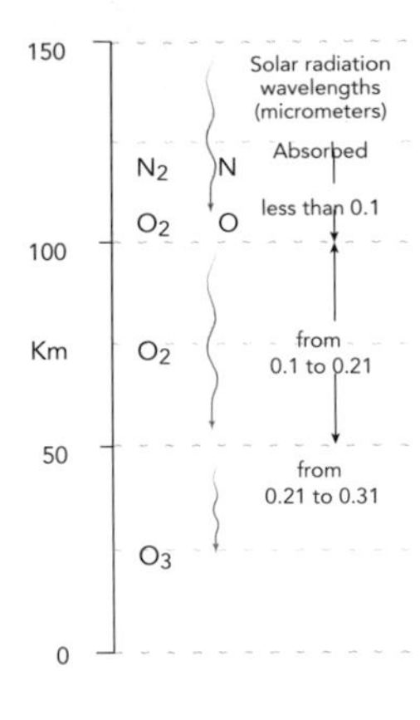

Absorption of short wave solar radiation in the high atmosphere

actually divides an O_2 molecule into two oxygen atoms (O + O). In a split second, these extremely reactive atoms will combine with other O_2 molecules to produce ozone.

Additionally, ozone is destroyed by being split into O_2 and O by ultra violet radiation with a wavelength longer than 0.29 micrometers. This process is known as photodissociation. Thus, stratospheric ozone is not only created naturally, but also destroyed by entirely natural processes. The absorption of the radiation warms the atmosphere within the ozone layer and also prevents this radiation, which is harmful to life, from reaching the Earth's surface or lower atmosphere.

Creation and destruction are continuous during sunlit hours and, until the recent problems associated with serious artificial depletion, occurred at about the same rate. This meant that the total amount of the gas remained broadly constant. That said, the balance can be tipped in one direction or the other by the seasonal cycle, by volcanic eruptions, by changes in the intensity of the Sun's output, and by the quasi-biennial oscillation which is a switch in the flow within the equatorial stratosphere, from easterly to westerly and back to easterly over a period of about 26 months.

Ozone warms

The heating of the air, caused by the reaction between solar radiation and ozone, is very important indeed, because it plays a part in determining the circulation of the atmosphere at high elevations and, to some extent, of the troposphere below. The reason for this is that pressure patterns are strongly influenced by thermal patterns; in their turn, wind direction and strength are determined by the horizontal pattern of pressure.

Where is ozone found?

Not surprisingly, stratospheric ozone is formed mainly within tropical latitudes, where solar radiation is strong throughout the year. However, this is not where the highest concentrations are to be found, as atmospheric circulation transports the ozone to extratropical latitudes.

Ozone values in the tropics are not only low because it is carried away from the region, but also because the stratosphere is shallower there than across higher latitudes. This is because the low-latitude troposphere is about 18 km [11 miles] deep, whereas it is about a third of that value in the highest latitudes. The largest amounts of ozone occur in middle latitudes, above Hudson's Bay and eastern Siberia in the northern hemisphere, and around the flank of the Antarctic continent in the southern. Concentrations decline toward polar regions, especially over Antarctica. Even in the stratosphere between 12–35 km [8–22 miles], an area of high concentration, it represents only about two parts

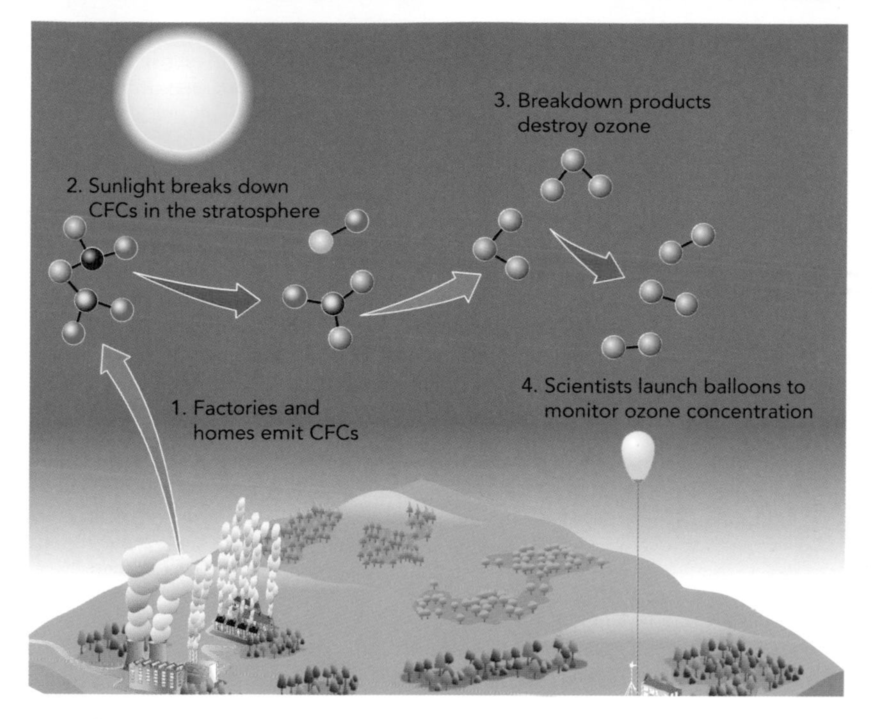

Process of ozone depletion due to the release of CFCs into the air

per million by weight. Therefore, it is a gas with a significance far exceeding its minute presence. Until only a couple of decades ago, there was no suspicion that there was any risk at all to this natural shield.

Supersonic transport

In the early 1970s, the thrust, on both sides of the North Atlantic, to construct and fly supersonic transport (SST) passenger aircraft caused some stratospheric specialists to voice concern. To achieve speeds of Mach 2, such aircraft were required to cruise in the lower stratosphere – the very region where ozone is most highly concentrated. The prime concern of the scientists was that reactive nitrogen in the SST exhausts might speed up the natural decay of ozone and lead to significant depletion. As it turned out, only a small number of Concorde SSTs were constructed, the first making its maiden commercial flight to the Gulf in 1976. Since then, the impact of these few aircraft has been insignificant.

Chlorofluorocarbons

Around the time that suspicions about SSTs surfaced, scientists also became concerned about the potential threat to the ozone layer from artificial compounds known as chlorofluorocarbons (CFCs). These were developed in the 1920s from chlorine, fluorine, and carbon, and have an extremely stable molecular arrangement. They are non-toxic, non-corrosive, non-flammable and do not react with almost any other substance. These properties led to their use as coolants in refrigerators and air conditioners, as trappers of heat in insulated cups and houses, as spray-can propellants, and as cleaners of delicate electronic components.

Once released into the air, however, the extreme stability of CFCs meant that they would drift around for years – not only near the surface, but also way up in the stratosphere. At these higher altitudes, the very short wave solar radiation would be able to split their bond to add millions of tons of extra chlorine to the stratosphere (this gas does occur there naturally, from dimethyl chloride supplied by the oceans).

In the same way that governments face tough decisions today about global warming, political leaders in the early 1970s had to decide whether the scientists' prediction would become a reality. If they believed the experts, massive changes would have to be implemented to slash CFC production and to find safe replacements. Dismissing the scientific assessment would be courting potential disaster on a global scale.

Unraveling the problem

So it was that in 1974, atmospheric scientists set out to learn as much as possible about the complexities of stratospheric chemistry, and about ozone in particular. One problem was that there were few observations of important chemicals in the stratosphere. Over the following few years, chemists experimented in laboratories to determine the rate at which chlorine could destroy ozone. Others launched special balloons that carried instrument packages to measure the concentrations of key chemicals that help control ozone levels.

By 1976, the message from many of the experts involved was that their observations and computer simulations pointed to a serious depletion of stratospheric ozone due to the release of CFCs. The public called for action from governments, but it was not until 1979 that some, including that of the largest producer and user – the United States – stopped the sale of aerosol cans that used CFCs as propellants. This action led to a quick leveling off of CFC production because spray-cans were the largest consumer of these gases.

The Vienna Convention

The problem persisted, however, because industry still used CFCs in other ways, and by 1985 production was increasing at a rate of 3% a year. This significant change stimulated governments to get together in that year to sign the Vienna Convention, which required the parties involved to formulate a plan for global action to curb these gases. Additionally, the scientists were required to present the latest information about ozone depletion due to CFCs and also halons, which are related bromine-based compounds. Use of the latter had increased dramatically during the previous decade because of their efficiency in extinguishing fires.

In the mid-1980s, the best estimate of the reduction in stratospheric ozone pointed to a decrease of some 5% by 2050, which would mean millions of new cases of skin cancer worldwide. This was serious and was compounded by the fact that the halocarbons already injected into the air would remain there for more than a century.

Discovery of the "hole"

In May 1985, news spread around the world that scientists had discovered massive ozone depletion over parts of the Antarctic during the austral spring. The reduction was so great that there were almost holes in the ozone layer, and the concept stuck.

In fact, there is no region of the stratosphere where there is no ozone, and the discovery in the mid-1980s pointed to severe depletion above the Antarctic alone. This was confirmed by balloon-borne measurements and satellite observations. There was no question that the depletion in the early spring was real – but was it really due to human society polluting the stratosphere or, perhaps, some natural variation?

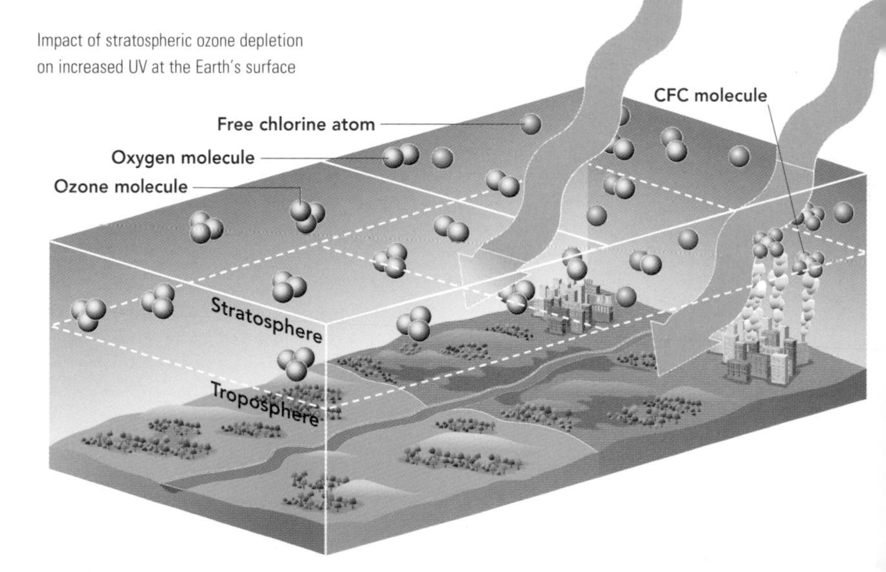

Impact of stratospheric ozone depletion on increased UV at the Earth's surface

Flying through the "hole"

Massive effort was put into unraveling the problem. In September 1986, a team of scientists traveled to McMurdo Station in the Antarctic, where they undertook an intensive programme of ground-based and balloon-borne observations. They quickly confirmed the presence of high levels of artificially-made, ozone-destroying chemicals.

One year later, in the subsequent austral spring, over 100 scientists made their way to Punta Arenas in southern Chile. Others returned to McMurdo to repeat their activities of the previous year. High-flying U-2 aircraft, carrying special instruments, flew through the intensely cold layers of the lower stratosphere, where depletion was known to occur. The experiment confirmed that chlorine and bromine pollution had indeed led to ozone depletion. The scientific suspicions of a decade before had been confirmed.

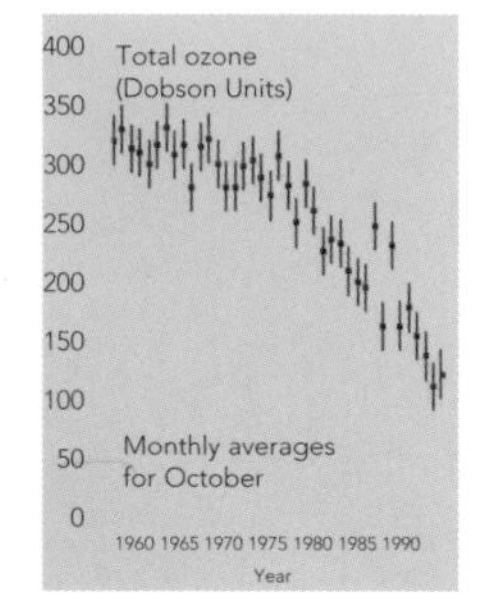

October average total ozone (1956–94), Halley Bay, Antarctica

Antarctic depletion explained

The north and south polar regions are distinctly different, being almost mirror images of each other, in that the Arctic is an ocean surrounded by continents, while the Antarctic is a continent surrounded by the circumpolar ocean. This fact plays a crucial role in the difference between their respective high-level (and low-level) circulations.

In the depths of winter, the stratospheric circulation above the Antarctic is a more-or-less smooth westerly flow centered near the pole. It is not particularly wavy since it is above the broadly symmetric Antarctic continent. This flow means that very frigid air resides inside this spinning well during the southern-hemisphere winter – often plummeting below –80°C [–176°F] – which leads to the formation of polar stratospheric clouds (PSC). Breakdown of ozone occurs when the gas interacts with the CFCs on the ice crystals forming these clouds. When springtime solar radiation starts to increase in September and October, the circulation breaks down, becoming wavy and letting in ozone-rich air from higher latitudes. This raises the concentration of gases to safer levels until the circulation cuts off again as the next winter approaches.

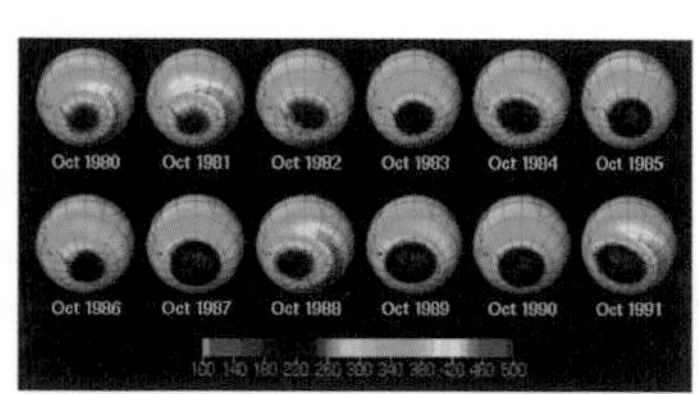

Southern hemisphere October mean ozone totals 1980–91

Arctic difference

The Arctic is quite different: it is much more susceptible to the incursion of frontal disturbances deep into its interior, especially through the Norwegian Sea. This means that, in most years, the stratosphere above it does not become cold enough for the formation of PSC and, therefore, does not suffer a similar annual depletion. On occasion, however, some unusually cold winters have led to ozone depletion within the Arctic stratosphere.

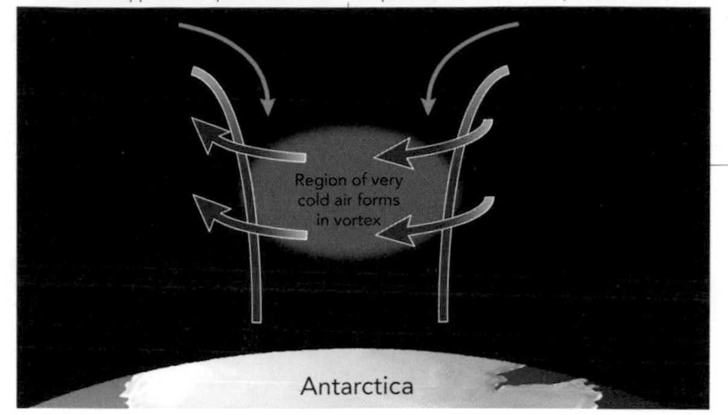

Antarctic polar vortex

The Montreal Protocol

The discovery of serious depletion over the Antarctic spurred governments into action. In September 1987, a meeting of environment ministers from 24 industrialized nations was convened in Montreal, Canada. They agreed a plan of action to reduce CFCs so that the 1986 concentrations would be reduced by 20% by the end of 1994, and by 50% by the end of 1999. This binding agreement was embodied in the Montreal Protocol.

The agreement represented a new step in intergovernmental collaboration on tackling environmental problems. Their continuing concern was expressed by their decision to reconvene in London, in 1990, to establish any necessary modification to the action taken.

The dramatic springtime ozone depletion persisted – and not only for the austral pole. The Arctic polar region was under intense study in the late 1980s. Depletion occurred there, too, but not with the same intensity.

The London Amendment

In June 1990, diplomats met in London, UK, to reinforce agreements made in Montreal. They declared a complete end to the production of CFCs by 2000, of halons (except for absolutely essential use) by 2000, and of carbon tetrachloride (by 2000) and methyl chloroform (by 2005). Included in this London Amendment were provisions for easier phase-outs by developing countries, and the establishment of a fund to help them switch to ozone-friendly replacements.

The Copenhagen Amendment

In 1992, the concerned parties gathered again in Copenhagen, Denmark, to bring deadlines forward. Thus, all CFC, carbon tetrachloride and methyl chloroform production was to cease by the end of 1995, and halon manufacture even earlier, by the end of 1993. Other harmful agents were to be phased out completely by the end of 2029.

Glossary

Absolute humidity Mass of water vapor contained in unit volume of air (including the water vapor)

Adiabatic cooling/warming Process of temperature change within ascending/descending air due to expansion/compression of the air. No heat is exchanged between the air and its "environment"

Advection Transport rate of heat, moisture, etc. by motion of the atmosphere

Advection fog Fog produced by the cooling of damp air as it moves across a cooler surface

Aerosol Microscopic liquid and solid particles suspended in the atmosphere

Airmass An extensive area of air that has broadly similar temperature and humidity characteristics

Albedo The fraction of radiation reflected by a surface (usually expressed as a percentage)

Altocumulus Middle level cumulus cloud

Altostratus Middle level stratus cloud

Anemogram Paper strip chart used in an anemograph

Anemograph Instrument that charts a continuous trace of wind speed and direction

Anemometer Instrument for measuring wind speed

Aneriod barometer Instrument that measures atmospheric pressure

Anticyclone An extensive area of high pressure

Azores high Region of high pressure that often occurs across the subtropical North Atlantic

Back Counterclockwise change of wind and time at one place or with height above one place at a given time

Barograph An instrument that measures atmospheric pressure and displays it as a time trace

Barometer An instrument that measures atmospheric pressure

Barometric pressure Another term for atmospheric pressure, as measured by a barometer

Beaufort Scale Means of assessing wind speed visually by relating it to how disturbed the sea surface is (or features of the land surface are)

Bergeron-Findeisen process Process in clouds whereby snowflakes grow at the expense of liquid water droplets where both are present

Blizzard A condition in which blowing snow seriously reduces surface visibility

Chloroflourocarbons (CFCs) Artificial chemical compounds that are powerful greenhouse gases and responsible for stratospheric ozone depletion

Cirrocumulus High cumulus cloud

Cirrostratus High stratus cloud

Cirrus High cloud with "wispy" or "striated" form

Cloud droplets Tiny liquid droplets that compose cloud. Typically 0.02 mm [0.0007 of an inch] diameter

Cloud street A line of shallow cumulus clouds that are generally non-precipitating, orientated with the wind shear between their base and top

Coalescence Process in clouds whereby droplets grow larger by "bumping" into each other

Col A region of light wind on a surface pressure map that lies between two highs and two lows

Cold front The leading edge of cold air

Condensation Deposition of liquid water or ice from water vapor onto small nuclei in the atmosphere

Condensation nucleus Found in every cloud droplet or ice crystal; critical microscopic particle on which condensation occurs

Convergence The process whereby air flows toward a line or "zone" from different directions

Conveyor belt Large-scale current that flows through a frontal depression

Cumulonimbus The deepest form of cumulus cloud, associated with precipitation, and thunder and lightning

Cumulus A "heap" cloud that varies from a few hundred meters deep to troposphere deep

Cut-off A near-circular wind pattern in mid-latitudes and the subtropics; normally colder air that has been isolated or cut-off from very meandering higher latitude flow

Cyclone A traveling low pressure disturbance often associated with wind, cloud, and precipitation

Depression A traveling low pressure disturbance usually associated with bad weather

Dew Deposition of liquid water droplets on the surface of the Earth by direct condensation from water vapor

Dewpoint temperature The temperature to which a sample of air must cool, at constant pressure and humidity mixing ratio, in order for saturation to occur

Diffuse radiation Part of solar radiation that comes from the sky, not directly from the solar "disk"

Direct radiation Part of solar radiation that comes directly from the solar "disk"

Divergence Process whereby the flow accelerates along its line of motion

Doppler radar Instrument used to map the component of air motion toward or away from itself, to spot low-level convergence/divergence, etc

Downdraught Localized, sudden, and often strong gust of cool air within a shallow surface layer. Originates at higher levels within a deep convective cloud, flowing out at the surface

Drizzle Very small droplets of liquid precipitation with diameters between 0.2–0.5 mm [0.007–0.02 inch]

Dry adiabatic lapse rate Rate at which an unsaturated parcel of air will cool/warm as it ascends/descends due to its expansion/compression. It is fixed at 9.8 C [50 F] per km

Dry bulb temperature Air temperature sensed by a mercury-in-glass thermometer

Dust storm Storm in which very large amounts of dust are raised upward many hundreds of meters by strong winds. Occur mainly in hot deserts

El Niño Occasional relatively rapid warming of the sea surface locally around Peru during December, and/or widely across the equatorial Pacific

Ensemble method Forecasting technique which involves running many predictions in parallel, each of which has virtually the same initial conditions save a subtle change that varies from one forecast to the next.

ENSO (El Niño Southern Oscillation) Occasional large-scale air-sea interaction across the tropical Pacific, associated with even larger-scale climate anomalies over a year or so.

Environmental lapse rate Rate of change of temperature (and humidity) with height

Evaporation Process whereby liquid water changes to vapor

Evaporative cooling Process whereby air, for example, can be cooled by the evaporation of water drops held within, or falling through, it.

Eye Region at the center of hurricanes, etc

Eyewall cloud "Cylinder"-like region of cumulonimbus cloud and torrential rain that surrounds the eye in hurricanes, typhoons, etc

F-scale Scale that links the damage produced by

tornadoes to their estimated wind speed

Ferrel cell The middle latitude cell in the vertical plane and north-south direction, with descent in the subtropics and ascent in the middle latitudes

Fog Horizontal visibility at the surface of less than 1000 m [3000 feet] due to suspended water droplets within the air

Front Leading edge of warm or cold air

Frontal depression Traveling low pressure disturbance with attendant warm, cold and, often, occluded fronts

Frost Condition of zero or sub-zero temperature often noted by icy deposit on grass, etc

Frostpoint The temperature at which air is saturated with respect to ice, by cooling at constant pressure and humidity mixing ratio

Funnel cloud Narrow snaking cloud that protrudes from the base of a parent cumulonimbus, associated with a tornado

Geostationary/geosynchronous Satellite orbit that is in phase with the speed of the Earth's rotation

Global warming The increase of global mean temperature believed to be associated with the artificial increase in greenhouse gas concentration

Graupel Ice particles from 2–5mm [0.08–0.2 of an inch] in diameter, formed by accretion in some clouds

Greenhouse gas A gas that partly absorbs outgoing Earth radiation and re-emits part of it back down to act as a kind of "insulator"

Gust front A line at the surface along which wind speed (and often direction) increases suddenly and strongly

Haboob Sand/dust storm in parts of Saharan Africa

Hadley cell The tropical cell in the vertical plane and north-south direction, with deep ascent at low latitudes and subsidence in the subtropics

Haze Reduced horizontal visibility associated with high concentrations of very small suspended solid particles leading to a whitish appearance to the sky

Heat low A surface low pressure area generated over land/continental regions by intense heating

Heterosphere Layer of the atmosphere, above about 100 km [60 miles], where gases are layered/separated

High An extensive area of relatively high pressure; can be slow-moving or mobile; generally associated with dry weather

Hill fog Fog caused by stratiform cloud intersecting hills

Hoar frost Icy deposits that form under generally calm, cloud-free conditions at night

Homosphere Region of the atmosphere in which the constituent gases (excluding water vapor) are well-mixed, and in virtually constant proportion

Humidity A measure of the water vapor content of air, expressed as relative humidity, absolute humidity, etc

Hurricane An intense cyclonic circulation that forms in the tropical North Atlantic and North-East Pacific.

Hygrogram Strip chart produced by a hygrograph

Hygrometer Instrument measuring humidity of air

Hygrometric tables Tables that relate values of dry bulb, wet bulb, and dewpoint temperature to each other, and to other humidity values

Ice fog Fog formed of ice crystals at low temperatures

Inversion A layer in the atmosphere within which temperature increases with height, at the surface on a clear, calm night, or above the tropopause, for example. Inversions of humidity also occur

Ionosphere Deep layer of the atmosphere above about 60 km [40 miles] in which the concentration of ions and free electrons reflects radio waves

Iridescence Patches of red and green, or sometimes blue and yellow, that occur on high clouds, most often within some 30 degrees of the Sun. Caused by the diffraction of sunlight by small cloud particles

Isobar Contour of constant mean-sea-level atmospheric pressure

Isotach Contour of constant wind speed

Isotherm Contour of constant temperature

Jetstream Well-defined zone of very strong winds

Kelvin Unit of temperature (K) with a base at absolute zero, such that 0C = 273K, 100C = 373K

Knot Unit used to express wind speed, one nautical mile per hour

Lapse rate The rate at which a variable changes with height in the atmosphere. A positive lapse rate means a decrease with increasing height.

Lee waves Waves in the troposphere often made visible by stationary lens-shaped clouds, to the lee of hills or mountains

Lenticular cloud Lens-shaped clouds that mark the presence of lee waves

Lightning Massive electrical discharge from cloud-to-cloud, cloud-to-ground or cloud-to-air, associated with thunderstorms

Long wave radiation Radiation emitted with relatively long wavelength (with respect to short wave solar radiation) by the Earth and atmosphere

Low An extensive area of relatively low pressure that is often mobile, but occasionally stationary

Maximum temperature The highest value of temperature recorded at one site over a fixed period of time, most commonly 24 hours

mbar Abbreviation for "millibar" or one thousandth of a bar. Unit of pressure.

Mean-sea-level pressure Atmospheric pressure measured at a site has to be "reduced" to mean-sea-level, and a "correction" must be applied to the majority of barometric readings not taken at sea-level

Mercury barometer "Official" means of measuring atmospheric pressure via the fluctuations in height of a mercury column

Mercury thermometer "Official" means of measuring dry bulb and maximum temperature via the thermal expansion of mercury

Mesopause Upper boundary of the mesosphere at about 80 km [50 miles] above sea-level

Mesosphere Layer of the atmosphere above the stratosphere characterized by a decrease of temperature with height. Stretches from about 50–80 km [30–50 miles] above sea-level

Minimum temperature Lowest dry bulb temperature measured in a fixed time, e.g. 24 hours

Mist Reduced horizontal visibility greater than 1000 m [3000 feet] due to suspended water droplets

Monsoon Very large, sub-continental, scale reversal of surface wind direction on a seasonal basis

Nimbostratus Deep, precipitating stratiform cloud

Occluded front/occlusion Front with warm air lifted off the surface, cool or cold air at low levels, cloud and precipitation

Orographic cirrus High-level cloud occasionally formed by flow over a mountain/high hill roughly at right angles to the upper tropospheric flow

Orographic cloud Cloud formed by condensation within moist air flowing over hills or mountains

Ozone The tri-atomic form of oxygen

Ozone depletion The reduction of ozone concentration in the stratosphere due to the presence of artificial constituents

Noctilucent cloud Thin bluish-white cloud around 80–90 km [50–60 miles] elevation, best viewed at twilight in polar and higher latitudes

Pitot tube Narrow tube used to measure wind speed by sensing pressure imposed by airflow on the one open end; used on aircraft, for example

Polar cell Weak circulating cell in the vertical north-south plane, stretching from polar regions to higher middle latitudes

Polar orbiter Type of satellite orbit that crosses above polar regions on every orbit

Precipitation Solid and liquid particles that fall/settle within the atmosphere

Pressure Measure of the downward force per unit area at the Earth's surface exerted by the atmosphere above a point, reduced to the datum of mean-sea-level

Pressure tendency Rate of change of barometric pressure at a site, usually over three, six or 24 hours

Quasi-biennial Oscillation Reversal of wind direction in the equatorial lower stratosphere from westerly to easterly to westerly with an average 26 month period

Radiation fog Fog formed through strong cooling by radiative losses at the surface/lower atmosphere

Radiative warming/cooling The process of changing the temperature of air, for example, by the absorption (heating) or emission (cooling) of radiation

Radiometer Instrument that measures the intensity of infrared radiation emitted by a body

Radiosonde Balloon-borne instrument package that senses and relays data on dry bulb temperature, relative humidity, pressure, and wind

Rainbow Optical phenomenon generated by refraction and internal reflection of sunlight shining onto a falling shower of raindrops, resulting in an arch of concentric colored bands in a spectral sequence

Raingage Instrument that measures equivalent depth of rain, as a total over a fixed period, or continuously

Relative humidity The ratio (expressed as a percentage) of the actual absolute humidity to the saturation value at the reported temperature

Ridge A region of high pressure that emanates from a larger anticyclone

Rime Frost formed by the deposition of supercooled water drops under windy conditions

Roaring Forties Mid-latitude region of the Southern Ocean across which winds often reach gale-force

Rossby wave Large-scale "long wave" in middle and upper troposphere in middle and higher latitudes

Saffir-Simpson scale Intensity scale relating surface wind speeds in hurricanes to their damage and surge height

Saturated adiabatic lapse rate Rate at which saturated air cools/warms due to expansion/compression as it ascends/descends

Saturation State of a parcel of air that contains the maximum possible amount of water vapor for its temperature

Screen A white wooden box housing standard thermometers, etc, at a weather observing site

Scud Ragged low cloud that moves rapidly in the wind below a higher deck of rain cloud

Sea fog Fog formed by the passage of warm, moist air blowing across a cooler sea

Short wave radiation Relatively small wavelength radiation emitted by the Sun

Shower A short-lived, intense period of rain, hail, or snow falling from a deep cumulus cloud

Snow Solid precipitation formed by the coagulation of ice crystals into various hexagonal shapes

Solar constant Magnitude of the constant flow of power from the Sun received at right angles to the solar beam at the top of the Earth's atmosphere equal to 1376 watts per square meter solar

Solarimeter Instrument that measures the intensity of solar radiation

Southern Oscillation A "see-saw" in mean-sea-level pressure; e.g. when the South-East Pacific High is weaker than average, the low to the north of Australia is shallower

Specific humidity Concentration of water vapor (gm) contained in a kilogram of air

Steam fog Shallow fog that forms as chilly air flows across much warmer water

Storm surge A positive departure in the elevation of the sea surface produced by "doming" underneath a traveling low pressure system

Stratiform General term for all layered cloud

Stratocumulus Low cloud that is sheet-like but composed of individual flattened "lumpy" cells

Stratosphere Layer above the troposphere characterized by very stable conditions and dry air

Stratus Low cloud that has a featureless, flat base

Subcloud layer Zone between the surface and cloud base

Subsidence The process of sinking of air

Subtropical anticyclones Semi-permanent features of global weather and climate situated over the subtropical oceans

Supercooled cloud Cloud composed of liquid water droplets at a temperature below zero Celsius

Synoptic chart Map that depicts various weather elements at the surface or in the upper air at a specific time

Temperature Measure of the heat content of air

Thermal A plume of relatively warm air that ascends invisibly through a cooler environment

Thermal advection Change of temperature over time at one spot due to the horizontal movement of cold and warm airmasses

Thermocline Layer in the ocean through which temperature declines rapidly, separating the upper-ocean mixed layer from the cold, deep water

Thermogram Strip chart used in thermographs

Thermograph Instrument that produces a time trace of dry bulb temperature, often for one week

Thermosphere High region of the atmosphere above the mesopause within which temperature increases with increasing height

Thunder Sound generated by intense and sudden heating of air by lightning

Thundercloud Deep convective cloud in the troposphere that is electrically active with lightning and therefore thunder

TIROS Television and Infrared Observation Satellite

Tornado Rapidly rotating, narrow, snaking column of air associated with a "parent" cumulonimbus. It must reach the ground and is often made visible by a funnel cloud

Trades The North-East and South-East winds that blow across the tropical oceans, converging into the ITCZ

Tropical maritime Type of airmass with a source in subtropical anticyclones over the sea. Characterized in middle latitudes by mild, humid air in the warm sector of depressions

Tropical storm Particular stage of a tropical cyclone when the system receives a name; category immediately below hurricane or equivalent

Tropopause "Lid" that caps the troposphere, thin layer or surface where the temperature decrease with height ceases

Troposphere The lowest layer of the atmosphere characterized by deep overturning motion and virtually all weather. It is deepest at low latitudes and shallowest at high latitudes and typified by decreasing temperature with increasing height

Trough An elongated region of low pressure, often with a long axis along which there is a marked cyclonic wind shift

Turbosphere Deep layer of the Earth's atmosphere within which the gaseous components of the dry atmosphere are well mixed

Turbulence Small-scale (but can be larger-scale) random fluctuations of windspeed and direction

Typhoon Regional name for a tropical cyclone that has reached hurricane-equivalent intensity in the North-West Pacific

Ultraviolet Region of the electromagnetic spectrum with wavelengths shorter than the violet end of the visible spectrum

Updraught Upward-flowing air within cumulus clouds and invisible thermals. They vary from a few to many tens of meters per second

Vane A vertical, plate-like indicator that points into the wind to reveal its direction

Vapor pressure That part of barometric pressure exerted by the column total of water vapor above a point

Veer A clockwise change of wind direction with time at one place OR with height above one place at a given time

Virgae Lines of precipitation that appear from cloudbase but do not reach the surface

Visibility The minimum horizontal visibility observed at a site at a particular time

Visible radiation The range of wavelengths of radiation to which our eyes are attuned

Walker cell Large-scale cells in the vertical plane around the equator

Warm front The leading edge of the warm, moist air in a frontal depression. A gently sloping surface

Warm sector The region between the warm and cold fronts, generally cloud-laden, mild and moist air with occasional precipitation

Water vapor Invisible form of water in the atmosphere

Wet bulb temperature The temperature recorded by the wet bulb thermometer in a screen; that to which air cools by evaporating water into it until saturated (at constant pressure)

Wind Air in motion; speed measured by an anemometer

Windshear Change of wind direction and/or speed above a site at a given time or in the horizontal plane

Zonal mean Produced by averaging all values of, say, mean surface temperature for a month/season. etc, at all sites in a latitude strip to obtain one value

Further Reading

Ahrens, C. D. **Meteorology Today: An introduction to weather, climate & the environment**, West Pub Co, 5th Edition 1994
Baker, K. **Be your own weather expert**, Merlin 1995
Barry R. G. & Chorley, R. J. **Atmosphere, Weather & Climate**, Routledge, 6th Edition 1992
Burroughs W. J. et al, **Collins Weather, The ultimate guide to the elements**, Harper Collins 1996
Crowder, Bob **The Wonders of the Weather Service**, Australian Govt Pub 1995
Dunlop, S. **Collins Gem Weather Photoguide**, Harper Collins 1996
Eden, P. **Weatherwise**, MacMillan 1995
Houghton, D. **Weather at Sea**, Fernhurst, 2nd Edition 1998
Houghton J. T. (Ed), **Climate Change 1995, The science of climate change**, CUP 1996
McIlveen, R. **Fundamentals of weather and climate**, Chapman & Hall, 3rd Edition 1992
Pedgley, D. **Mountain Weather**, Cicerone Press, 2nd Edition 1997
Williams, J. **The Weather Book**, USA Today 1992

Acknowledgements

Broomfield C. S. 119 bottom right
© Crown. Reproduced with the permission of the Controller of HMSO 23, 26, 36 top and bottom, 40, 46, 90
© EUMETSAT 1999 42, 43, 44, 45, 59, 65, 66, 68
Galvin J. F. P. 34 center left, bottom left, bottom center, 56, 103 top
Goddard B. 36/37
Hubert S. 35 bottom right, 103 bottom left, 107, 108, 116, 118 bottom
Hulbert J. E. L. 27, 32
Image Colour Library Limited 145
Maton G. 106 right
McConnell D. 21, 34/35 center top
NOAA 129, 130
Nye P. 15
Pouncy F. J. 25
Reynolds R. 34/35 bottom center, 35 top right, center right, 39, 85, 102, 118 center
Science Photo Library NASA/Science Photo Library 5 and 16, Claude Nuridsany & Marie Perennou/Science Photo Library 105, David Frazier/Science Photo Library 118 top, Larry Miller/Science Photo Library 121 and 138
Shine K. P. 119 top right
Telegraph Colour Library Darryl Torckler 34 top left, J. L. Manaud 91 left, Michael Kraft 91 right, Andrew Mounter 117 top, Marx 117 bottom
Tony Stone Images Schafer & Hill 34/35 center, Ron Sanford 81 and 100, Hans Peter Merten 99, Randy Wells 117 center, David Woodfall 119 top left, David James 119 top center, John Lamb 119 center right, Frank Oberle 165
University of Dundee 41 right, 47, 57, 89, 93, 97 top and bottom, 101, 142
Walton J. A. 87
Watts P. D. 103 bottom right
Young M. V. 106 left
Young W. K. 24

The illustrations have been prepared by Julian Baker and Stefan Chabluk.
Some artwork has been produced using reference material from the following sources:

Cambridge University, Department of Physical Chemistry 181
Cambridge University Press 185
Crown.141 right
Crown.142 right
Hadley Center 175
Hadley Center 176 bottom
Hadley Center 177
Hadley Center 178
NCAR (National Center for Atmospheric Research) 133 bottom
NOAA (Atlantic Oceanographic and Meteorological Library) 132-133 top
NOAA (National Center for Environmental Prediction) 170
NOAA (National Center for Environmental Prediction) 171
NOAA (National Center for Environmental Prediction) 173
NOAA (Pacific Marine Environmental Lab) 169
NOAA (Reports to the Nation) 183

Index

Page numbers in italic refer to illustrations/photography or their captions.
Page numbers in heavier type refer to main description